Oxford **Mathematics** 5
Primary Years Programme

Contents

OXFORD
UNIVERSITY PRESS
AUSTRALIA & NEW ZEALAND

Place value

In a number, the value of each digit depends on its position, or place.

923856 is easier to read if we write it as 923 856. It also makes it easier to say the number: nine hundred and twenty-three thousand, eight hundred and fifty-six.

I am worth 3 thousand

Guided practice

1 Look at this number: 725 384. The 7 is worth 700 000.
Show the value of the other digits on the place value grid.

	Hundred thousands	Ten thousands	Thousands	Hundreds	Tens	Ones	Write the number, using gaps if necessary
e.g.	7	0	0	0	0	0	700 000
a							
b							
c							
d							
e							

Remember to use a zero as a space-filler.

2 If we write thirty-two thousand, five hundred and nine in numerals, we use a zero to show there are no tens: **32 509**

Write as digits:

a   nine thousand, three hundred and seven   _____

b   twenty-five thousand and forty-six   _____

c   one hundred and two thousand, seven hundred and one   _____

- -

3 Write in words:

a   2860   _____

b   13 465 _____

c   28 705 _____

OXFORD UNIVERSITY PRESS

Independent practice

1. What is the value of the red digit in each number?

e.g. **85 306:** **80 000**

a 53 207: ____________

b 48 005: ____________

c 29 425: ____________

d 135 284: ____________

e 399 517: ____________

2. Write each number from question 1 in words.

e.g. 85 306: eighty-five thousand, three hundred and six

a ____________

b ____________

c ____________

d ____________

e ____________

3. Write these numbers as numerals.

a eighty-six thousand, two hundred and thirty-one ____________

b one hundred and forty-two thousand ____________

c six hundred and fifty-six thousand, three hundred and eight ____________

d one hundred and five thousand, nine hundred and twenty-one ____________

4. Circle the number that is **one more than** 25 789.

25 800 25 780 25 799 25 790

5 Expand these numbers. The first one has been done for you.

14 217:   10 000 + 4000 + 200 + 10 + 7

Remember to use spaces between the digits where necessary.

a   25 123:   20 000 + _____

b   63 382:   _____

c   6004:   _____

d   125 381:   _____

e   860 094:   _____

---

6 Use the digits on the cards to make:

| 6 | 1 | 5 | 3 | 9 | 7 |

a   the **largest** number using all the cards. _____

b   the **smallest** number if "5" is in the ones place. _____

c   the **largest** number if the "7" is in the hundreds of thousands place.

_____

d   the **smallest** number if the "1" is in the thousands place. _____

---

7 Write the number shown on each spike abacus as numerals and in words.

a
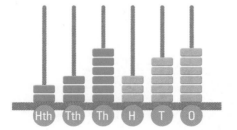
Hth Tth Th H T O

b
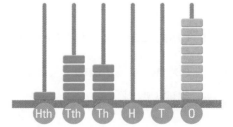
Hth Tth Th H T O

numeral: _____

words: _____

_____

numeral: _____

words: _____

_____

OXFORD UNIVERSITY PRESS

Extended practice

1 This table shows unusual record-breaking activities.

Place	Activity	Number
USA	Number of dogs on a dog walk together	
Spain	People salsa dancing together	
Poland	People ringing bells together	
Hong Kong	People playing percussion instruments together	
Singapore	People line dancing together	
Portugal	People making a human advertising sign	
Mexico	People doing aerobics at the same time	
India	Trees planted by a group in one day	
USA	People in a conga line	
England	The longest scarf ever knitted (in centimetres)	

Complete the number column in the table by rewriting the numbers below in order, from the **lowest** to the **highest** number. The events are in order from low to high.

Record numbers									
80 241	10 021	119 986	38 633	322 000	3117	34 309	3868	11 967	10 102

- -

2 The following numbers are from the list in question 1. They have been rounded in various ways. Write the actual number for each.

a   80 000   _____   f   10 000   _____

b   40 000   _____   g   100 000   _____

c   3000   _____   h   12 000   _____

d   300 000   _____   i   4000   _____

e   10 100   _____   j   35 000   _____

- -

3 Rounded to the nearest ten thousand, the 2006 population of Noosa in Queensland was 50 000 people. The actual number can be made by using each of these digits once: **1   2   5   6   9**

List as many of the 12 numbers that could be the actual population as you can.

_____

Finding a short cut

Imagine you were on a TV quiz show and had 4 seconds to answer the question. There are several strategies you could use to come up with the right answer. However, in only 4 seconds you would probably have to use a mental strategy.

For $100: What is 250 + 252?

Guided practice

1 You could use the **near-doubles** strategy for 252 + 250:
Double 250 is 500. Then add 2 = 502. Fill in the gaps.

	Problem	Find a near-double	Now I need to:	Answer
e.g.	252 + 250	250 + 250 = 500	add 2 more	502
a	150 + 160	150 + 150 =	add 10 more	
b	126 + 126	125 +		
c	1400 + 1450			

2 You could **split** the numbers. For example, 250 + 252 is the same as: 200 + 50 + 2 + 200 + 50. Fill in the gaps.

	Problem	Expand the numbers	Join the partners	Answer
e.g.	252 + 250	200 + 50 + 2 + 200 + 50	200 + 200 + 50 + 50 + 2 = 500 + 2	502
a	66 + 34	60 + 6 + 30 + 4	60 + 30 + 6 + 4 = 90 + 10	
b	140 + 230	100 + 40 + 200 + 30	100 + 200 + 40 + 30 = 300 + 70	
c	1250 + 2347			

3 You could use the **jump strategy** on an empty number line:

e.g. What is 50 + 52?

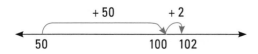

+ 50   + 2

50   100 102

Answer: 50 + 52 = 102

a What is 105 + 84?

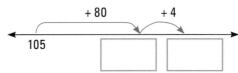

+ 80   + 4

105

Answer: 105 + 84 = _____

b What is 1158 + 130?

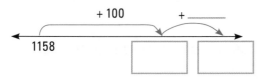

+ 100   + _____

1158

Answer: 1158 + 130 = _____

c What is 2424 + 505?

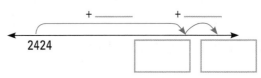

+ _____   + _____

2424

Answer: 2424 + 505 = _____

OXFORD UNIVERSITY PRESS

Independent practice

1 Another mental strategy for adding is the compensation strategy. It uses **rounding**. For 74 + 19, we can round 19 to 20 and say 74 + 20. Use the compensation strategy to solve these.

	Problem	Using rounding it becomes:	Now I need to:	Answer
e.g.	74 + 19	74 + 20 = 94	take away 1	93
a	56 + 41	56 + 40 = 96	add 1	
b	25 + 69	25 + 70 = 95	take away 1	
c	125 + 62	125 + 60 = 185	add	
d	136 + 198	136 +		
e	195 + 249			
f	1238 + 501			
g	1645 + 1998			

----

2 Use the compensation strategy to solve these.

a   35 + 99   _____

b   24 + 101   _____

c   173 + 198   _____

d   1407 + 1002   _____

e   1451 + 1499   _____

f   1562 + 1004   _____

----

3 Use the jump strategy to solve these.

a   125 + 38 =   ⟵————————————————————⟶

b   164 + 47 =   ⟵————————————————————⟶

c   1193 + 842 =   ⟵————————————————————⟶

d   2585 + 1321 =   ⟵————————————————————⟶

4 Practise the split strategy with these addition problems.

	Problem	Expand the numbers	Join the partners	Answer
e.g.	125 + 132	100 + 20 + 5 + 100 + 30 + 2	100 + 100 + 20 + 30 + 5 + 2	257
a	173 + 125			
b	1240 + 2130			
c	5125 + 1234			
d	7114 + 2365			
e	2564 + 4236			

5 Use your choice of strategy to find the answer. Be ready to explain the strategy you used.

a 713 + 190 = _____

b 1490 + 1490 = _____

c 2009 + 2009 + 2009 = _____

d 1864 + 3134 = _____

e 2499 + 1002 = _____

f 1236 + 247 = _____

g 2499 + 2499 = _____

h 3130 + 2360 = _____

OXFORD UNIVERSITY PRESS

Extended practice

Improving your estimating and rounding skills can help you save time with mental calculations.

1 Look at these facts and figures. Show how you would round the numbers by underlining or highlighting one of the numbers.

	World fact	Metres	Rounded number
a	Krubera: the deepest cave in the world	2191 m	2100 or 2200?
b	Cehi: the tenth-deepest cave in the world	1502 m	1500 or 1600?
c	Mont Blanc: the highest mountain in Europe	4807 m	4800 or 4900?
d	Mont Maudit: the tenth-highest mountain in Europe	4466 m	4400 or 4500?
e	Mt Everest: the highest mountain in the world	8850 m	8800 or 8900?
f	Mt Kosciusko: the highest mountain in Australia	2228 m	2200 or 2300?
g	Mammoth Cave: the longest cave in the world.	590 600 m	500 000 or 600 000?
h	Wind Cave: the fourth-longest cave in the world	212 500 m	200 000 or 300 000?

2 Circle the number that will make the information correct.

a The total of the depths of Krubera and Cehi caves is about

3500 m, 3700 m, 3600 m, 3400 m.

b Mont Blanc is about **20 m, 200 m, 30 m, 300 m** taller than Mont Maudit.

c If you walked the lengths of the Mammoth Cave and the Wind Cave you would

have travelled about **700 km, 70 km, 80 km, 800 km**.

3 Sarah goes shopping in a bargain shop. She has $11 to spend. She goes to the checkout with these items:

Paint set: $1.99	Ball: 99c	Calculator: $1.99	Cuddly toy: $1.99
Pen set: $1.25	Notebook: 49c	Geometry set: $1.99	Stickers: $1.29

a To the nearest dollar, how much more than $11 is the total?

b Which item should Sarah put back to be closest to a total of $11? _____

Addition written strategies

T	O
3	4
+ 2	5
5	9

One of the most common written strategies for addition is to set the numbers out vertically. You start with the ones and add each column in turn.

Sometimes you need to trade from one column to the next.

T	O
¹3	8
+ 2	5
6	3

Guided practice

1 Complete the following.

a

T	O
2	6
+ 2	3

b

H	T	O
1	3	3
+ 1	4	1

c

H	T	O
3	7	5
+ 1	2	3

d

Th	H	T	O
3	6	4	1
+ 1	2	2	5

2 Complete the following.

a

T	O
¹5	7
+ 2	9

b

H	T	O
1	¹2	8
+ 1	5	6

c

H	T	O
1	¹3	9
+ 2	8	6

d

H	T	O
6	6	8
+ 2	4	9

> You need to trade with these.

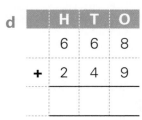

3 Start with the ones and add each column in turn.

a

H	T	O
2	4	9
+ 1	3	7

b

Th	H	T	O
3	2	4	6
+ 1	3	7	7

c

Tth	Th	H	T	O
3	2	2	8	6
+ 1	5	5	3	7

d

Tth	Th	H	T	O
4	2	7	4	2
+ 3	2	3	7	8

e

Hth	Tth	Th	H	T	O
4	3	4	5	3	6
+ 2	6	5	5	9	5

OXFORD UNIVERSITY PRESS

Independent practice

1 Look for a pattern in the answers for each row.

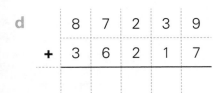

a
```
    8 5
+   3 8
———————
```

b
```
    5 3 8
+   6 9 6
—————————
```

c
```
    7 0 6 6
+   5 2 7 9
———————————
```

d
```
    8 7 2 3 9
+   3 6 2 1 7
—————————————
```

e
```
    6 2
+   5 9
———————
```

f
```
    1 5 8 9
+     7 4 3
———————————
```

g
```
    1 5 0 7 8
+   1 9 4 6 5
—————————————
```

h
```
    2 4 8 9 3 6
+   2 0 7 7 1 8
———————————————
```

i
```
    7 2
+   3 9
———————
```

j
```
      9 2 4
+   1 2 9 8
———————————
```

k
```
    1 8 6 5 1
+   1 4 6 8 2
—————————————
```

l
```
    1 8 6 1 2 8
+   2 5 8 3 1 6
———————————————
```

2 Look for linking numbers to save time in written addition.

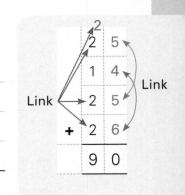

a
```
    2 7
    2 2
    2 3
+   1 8
———————
```

b
```
    2 1 4
    1 3 1
    1 9 6
+   2 7 9
—————————
```

c
```
    1 8 4
    2 3 5
    2 2 6
+   1 7 0
—————————
```

d
```
    4 7 5
    1 0 1
    1 3 5
+   6 0 9
—————————
```

e
```
    5 9 3
    2 1 8
    8 9 8
+   5 9 8
—————————
```

3

On a holiday, Jack spent $295 on food, $207 on travel, $985 for his hotel, $92 on presents and $213 on entertainment. He wanted to know how much he had spent and used a calculator and found that the total was $1612.

a   If you round the numbers, is Jack's answer reasonable?

b   How much did Jack spend altogether?

When you write an addition problem vertically, it is important to keep the digits in the correct columns. If you don't, you will get the wrong answer.

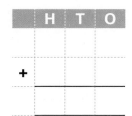

	T	O
	4	5
+	3	7
	4	8	7

✗

		T	O
		¹4	5
+		3	7
		8	2

✓

4 Rewrite these problems vertically, then solve them.

a   114 + 137

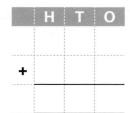

	H	T	O
+			

b   927 + 138

	H	T	O
+			

c   739 + 278

Th	H	T	O
+			

d   173 + 33 + 38

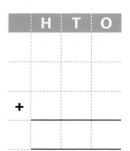

	H	T	O
+			

e   554 + 537 + 49

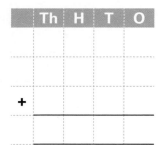

Th	H	T	O
+			

f   637 + 77 + 829

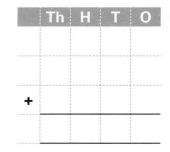

Th	H	T	O
+			

g   1452 + 257 + 2318

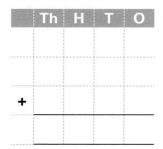

Th	H	T	O
+			

h   35 174 + 257 + 2318 + 624

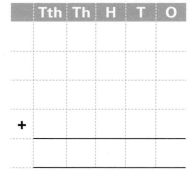

Tth	Th	H	T	O
+				

i   61 286 + 435 + 24 + 325

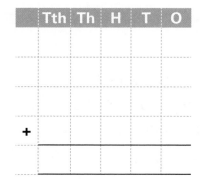

Tth	Th	H	T	O
+				

j   579 + 4529 + 33 + 6589 + 527

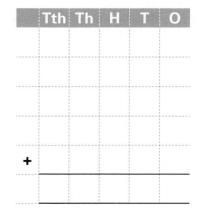

Tth	Th	H	T	O
+				

Extended practice

1 Find four different solutions to make this addition correct.

a

```
  3 □ 9
+   □ 6
─────────
□ 3 □
```

b

```
  3 □ 9
+   □ 6
─────────
□ 3 □
```

c

```
  3 □ 9
+   □ 6
─────────
□ 3 □
```

d

```
  3 □ 9
+   □ 6
─────────
□ 3 □
```

2 A football team can have more than 200 000 spectators at their home games in a season.

Here is some information about one famous football team.

- Number of home games: 12.
- Total number of spectators: 212 052.
- Average attendance per home game: 17 671.
- Every game had more than 10 000 spectators.
- No games had exactly the same number of spectators.

List the possible number of spectators for each game. Make sure the total is 212 052. Use the grid to help you keep the numbers in columns.

Game	Possible number
1	
2	
3	
4	
5	
6	
7	
8	
9	
10	
11	
12	
Total	

3 Find the total of 30 521 + 85 365 + 7570 and you will see that the digits in the answer make a pattern. Make three other three-line addition problems with the same answer.

Working-out space

UNIT 1: TOPIC 4
Subtraction mental strategies

Can you work out the answer to 76 – 19 in your head?

Round numbers are easier to work with.

We could say 76 – 20 instead of 76 – 19.

76 – 20 = 56. We took away 1 too many, so we add 1 back to the answer.

So, 76 – 19 = 57

Guided practice

1 Use the compensation strategy (rounding) to solve these. Fill in the gaps.

	Problem	Using rounding, it becomes:	Now I need to:	Answer
e.g.	76 – 19	76 – 20 = 56	add 1 back	
a	53 – 21	53 – 20 = 33	take away 1 more	
b	85 – 28	85 – 30 = 55	add 2 back	
c	167 – 22	167 – 20 = 147	take away ___ more	
d	146 – 198	346 –		
e	1787 – 390			
f	5840 – 3100			
g	6178 – 3995			

Splitting numbers can make subtraction easier. For example, 479 – 135 = ?

- Split (expand) the number you are taking away: 135 becomes 100 + 30 and 5
- First take away 100: 479 – 100 = 379
- Next take away 30: 379 – 30 = 349
- Then take away 5: 349 – 5 = 344
- So, 479 – 135 = 344

2 Use the split strategy. Fill in the gaps.

	Problem	Expand the number	Take away the 1st part	Take away the 2nd part	Take away the 3rd part	Answer
e.g.	479 – 135	135 = 100 + 30 + 5	479 – 100 = 379	379 – 30 = 349	349 – 5 = 344	344
a	257 – 126	126 = 100 + 20 + 6	257 – 100 =			
b	548 – 224	224 =				
c	765 – 442					
d	878 – 236					
e	999 – 753					

OXFORD UNIVERSITY PRESS

1 Use the **compensation** strategy to solve these—or find your own sensible short cut.

a 47 – 22 _____

b 184 – 29 _____

c 547 – 231 _____

d 2455 – 1219 _____

e 5667 – 2421 _____

2 Use the **split** strategy to solve these—or find another short cut.

a 45 – 24 _____

b 464 – 343 _____

c 676 – 254 _____

d 5727 – 3325 _____

e 8958 – 5635 _____

3 The split strategy can be used on an open number line. Fill in the gaps.

e.g. **What is 900 – 350?**

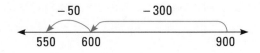

Answer: 900 – 350 = 550

a What is 776 – 423?

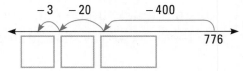

Answer: 776 – 423 = _____

b What is 487 – 264?

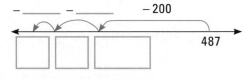

Answer: 487 – 264 = _____

c What is 1659 – 536?

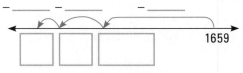

Answer: 1659 – 536 = _____

Another strategy for subtraction is to count up.

Tina buys a sandwich for $3.75. She gives a $5 note. To work out the change, the shopkeeper starts at $3.75 and counts up to $5.

$3.75 up to $3.80 is 5c

$3.80 up to $4 is another 20c

$4 up to $5 is $1

The change is 5c + 20c + $1 = $1.25

The difference between $3.75 and $5 is $1.25.

That is another way of saying $5 − $3.75 = $1.25.

4 Use the **counting-up** strategy to find the change if you paid for each item with a $10.

a   a toy at $7.50 _____

b   a book at $8.75 _____

c   a melon at $3.50 _____

d   a calculator at $4.45 _____

e   a game at $5.35 _____

f   a pencil set at $2.15 _____

You can also use the counting-up strategy to find the difference between ordinary numbers. For example, what is the difference between 200 and 155?

- 155 up to 160 is **5**
- 160 up to 200 is **40**

Altogether I counted up 45, so the difference between 200 and 155 is 45.

5 Use the counting-up strategy to work out the difference between these numbers.

a   100 − 57 = _____

b   150 − 128 = _____

c   200 − 135 = _____

d   151 − 118 = _____

e   1005 − 890 = _____

f   2500 − 2390 = _____

6 Use a mental strategy of your choice to find the answers to these problems. Be ready to explain the strategies you use.

a   89 − 19 = _____

b   65 − 14 = _____

c   78 − 21 = _____

d   150 − 75 = _____

e   1515 − 1220 = _____

f   2000 − 1450 = _____

OXFORD UNIVERSITY PRESS

Extended practice

1. A football game starts at 1:30 pm and ends at 3:05 pm. How long does it last?

2. The difference between two 3-digit numbers is 57. What might the numbers be?

3. Iva receives \$2.45 change after paying with a note. Which banknote might have been used and how much was spent?

4. What is 4235 – 397? Explain how you got the answer.

5. Bob, Bill and Ben buy the same model of car from different dealers.

   Bob pays \$7464 for his car. Bill pays \$193 more than Bob, but Bill pays \$193 less than Ben.

   How much do Bill and Ben pay for their cars?

6. Fill in the gaps to show three more ways to make the subtractions correct.

e.g.	6	1	3	–	5	3	5	=	7	8
a	6		3	–			5	=	7	8
b	6		3	–			5	=	7	8
c	6		3	–			5	=	7	8

Subtraction written strategies

Some written subtractions involve trading. Here is a reminder of how it works, using MAB and small numbers, such as 54 − 25.

When you write the algorithm, you trade in the same way.

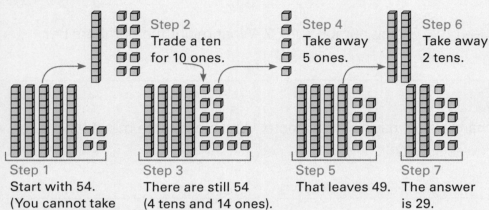

Step 2
Trade a ten for 10 ones.

Step 4
Take away 5 ones.

Step 6
Take away 2 tens.

Step 1
Start with 54. (You cannot take away 5 ones.)

Step 3
There are still 54 (4 tens and 14 ones).

Step 5
That leaves 49.

Step 7
The answer is 29.

There aren't enough ones. Trade a ten. That leaves 4 tens.

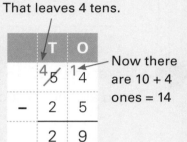

Now there are 10 + 4 ones = 14

	T	O
	⁴5̶	¹4
−	2	5
	2	9

Guided practice

1 You could use MAB to help with the trading as you complete these algorithms.

a

T	O
7	3
− 2	4

b

H	T	O
2	³4̶	¹3
− 1	2	7

c

H	T	O
4	5̶	4
− 2	3	5

d

H	T	O
7	2	5
− 3	1	8

e

Th	H	T	O
7	2	7	3
− 1	1	4	7

f

Th	H	T	O
4	3	6	1
− 1	2	6	7

g

Th	H	T	O
5	2	5	3
− 3	7	4	7

h

Th	H	T	O
6	7	7	1
− 2	7	7	3

i

Tth	Th	H	T	O
8	3	4	1	9
− 6	1	2	3	2

j

Tth	Th	H	T	O
4	3	7	2	4
− 2	5	4	6	5

k

Tth	Th	H	T	O
7	0	7	3	5
− 3	7	4	8	8

l

Hth	Tth	Th	H	T	O
8	1	3	5	1	8
− 2	4	5	8	7	9

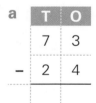

Subtraction with larger numbers works in the same way. Start with the ones column.

OXFORD UNIVERSITY PRESS

Independent practice

1 Practise trading with subtraction. Look for patterns in the answers.

a
H	T	O
4	1	0
−		8	9

b
H	T	O
5	0	8
−		7	6

c
H	T	O
8	1	2
−	2	6	9

d
H	T	O
8	7	2
−	2	1	8

e
H	T	O
9	5	3
−	1	8	8

2 Once you know how to subtract with trading, it doesn't matter how large the numbers are. Complete the following. There is a pattern in the answers.

a
Th	H	T	O
1	8	2	1
−		5	8	7

b
Th	H	T	O
3	7	1	4
−	1	3	6	9

c
Th	H	T	O
5	6	4	3
−	2	1	8	7

d
Th	H	T	O
6	1	5	5
−	1	5	8	8

e
Th	H	T	O
8	3	2	6
−	2	6	4	8

f
Th	H	T	O
8	4	1	3
−	1	6	2	4

g
Th	H	T	O
9	9	3	4
−			5	8

h
Th	H	T	O
9	7	5	3
−		9	8	8

3 There is also a pattern in the answers to these 5-digit subtractions.

a
Tth	Th	H	T	O
1	3	4	6	5
−		2	3	5	4

b
Tth	Th	H	T	O
2	6	0	8	1
−		3	8	5	9

c
Tth	Th	H	T	O
3	8	9	8	1
−		5	6	4	8

d
Tth	Th	H	T	O
6	6	1	3	3
−	2	1	6	8	9

e
Tth	Th	H	T	O
7	7	2	4	1
−	2	1	6	8	6

f
Tth	Th	H	T	O
9	1	2	3	5
−	2	4	5	6	9

4 Use the digits 1, 3, 2, 6, 4 and 7. Make the largest number using all the digits and the smallest number using all the digits.
Find the difference between the two numbers.

5 **Rounding** and **estimating** can help you avoid making careless mistakes. Imagine you subtract 189 from 913 and get an answer of 824. If you round and estimate you know the answer is wrong. 900 − 200 = 700, so the answer must be around 700. Write an algorithm and find the exact answer.

Working-out space

6 One algorithm in each pair is wrong. Estimate the answer, then circle the correct algorithm.

a
```
    6 1 2              6 1 2
  - 4 8 8     OR    - 4 8 8
  ───────           ───────
    1 2 4              2 2 4
```

b
```
    9 1 5 2            9 1 5 2
  - 2 9 5 8   OR    - 2 9 5 8
  ─────────         ─────────
    7 1 9 4            6 1 9 4
```

c
```
  1 4 2 0 5          1 4 2 0 5
  -   6 9 4 7  OR   -   6 9 4 7
  ─────────         ─────────
      7 2 5 8            8 2 5 8
```

Sometimes when you trade, there is nothing in the next column. Here's what to do:

More ones are needed ...

... but there are no tens

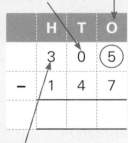

... so trade FROM the hundreds TO the tens first.

Trade a hundred.
That leaves 2 hundreds.

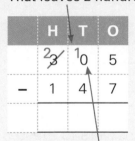

Now there are 10 tens.

Trade a ten.
That leaves 9 tens.

Now there are 15 ones.

7 Practise trading across two columns with these subtractions.

a
H	T	O
4	0	2
− 1	3	4

b
H	T	O
5	0	6
− 2	4	8

c
H	T	O
6	0	2
− 1	7	7

d
H	T	O
4	0	6
− 2	5	8

e
H	T	O
9	0	3
− 5	3	4

f
Th	H	T	O
3	4	0	7
− 2	5	8	9

g
Tth	Th	H	T	O
2	6	0	5	9
− 1	2	3	8	2

h
Hth	Tth	Th	H	T	O
5	3	0	7	7	2
− 1	4	4	8	4	6

OXFORD UNIVERSITY PRESS

Extended practice

1 Follow the rules to write three subtraction algorithms. Each algorithm must:

- be 5 digits take away 5 digits
- be different from the other two
- have the answer 999.

2 This table shows the size of the crowd at some sporting events around the world. Use the information to answer the questions.

Sport	Size of crowd	Year	Place	Working-out space
Gaelic Football	90 556	1961	Dublin, Ireland	
Hurling	84 865	1954	Dublin, Ireland	
Australian football	121 696	1970	Melbourne, Australia	
Rugby Union	109 874	2000	Sydney, Australia	
NFL	102 368	1957	Los Angeles, USA	

a What is the difference between the biggest and smallest crowds in the table? _____

b By how many was the American Football (NFL) crowd bigger than the Gaelic Football crowd? _____

c What is the difference between the total of the two Irish games and the total of the two Australian games? _____

d Use rounding strategies to circle the correct response.

The difference between the size of the crowds at the Rugby Union and Hurling games was about: **22 000**   **23 000**   **24 000**   **25 000**

3 The world's smallest dog is a Yorkshire terrier. It is only 76 mm from the ground to its shoulder. The tallest dog is a great dane, which measures 1054 mm high from the ground to its shoulder.

If they were side by side, what would be the difference in their heights? _____

The Ten Trick

Multiplying by ten is easy—but you don't just add a zero. The digits move one place bigger.

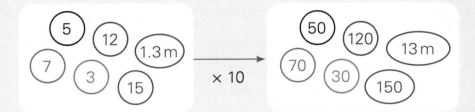

(If you added a zero to multiply 1.3 m by 10, the answer would be 1.30 m and that's the same length. It is clearly not the product of 1.3 m and 10.)

Guided practice

1 Complete the grid.

	e.g.		a		b		c		d		e			f		
	T	O	T	O	T	O	T	O	T	O	H	T	O	H	T	O
		4		7		8		6		9		1	4		1	9
× 10	4	0														

2 Multiply each of these by 10.

Remember, move the digits one place to the left when you multiply by ten.

14 m x 10 = ?
1 . 4 ☞
1 4

a   1.5 m _____

b   2.2 L _____   c   4.5 t _____

d $1.70 _____   e   3.8 cm _____

f   3.6 m _____   g $2.75 _____

When you multiply by 100, the digits move **two** places to the left. **For example, what is 11 × 100?**

Th	H	T	O
		1	1
1	1	0	0

3 Multiply by 100.

a   14   _____   b   17   _____

c   13   _____   d   27   _____

e   23   _____   f   45   _____

g   64   _____   h   3.7 m _____

i $1.25 _____

OXFORD UNIVERSITY PRESS

Once you know the ten trick, you can use it to multiply by multiples of 10.

In 5×30, 30 is the same as 3 tens, so change it to 5×3 tens.

$5 \times 3 = 15$ so 5×3 tens $= 15$ tens, or 150.

1 Fill in the gaps.

		× 20 Rewrite the problem and solve	**× 30** Rewrite the problem and solve
a	6	$6 \times 20 = 6 \times 2$ tens 6×2 tens $= 12$ tens 12 tens $= 120$ So, $6 \times 20 = 120$	$6 \times 30 = 6 \times 3$ tens 6×3 tens $=$
b	9		
c	8		
d	7		

Doubling

To multiply by 4, you can double a number and then double it again.

To multiply by 8, you can double a number three times.

2 Fill in the gaps.

		a	b	c	d	e	
×	**8**	**5**	**12**	**15**	**50**	**40**	**Strategy**
2	16						Double
4	32						Double again
8	64						Double again

Doubling and halving

If you double one number and halve the other, it can make multiplication easier.
It works like this: Imagine you didn't know that $5 \times 6 = 30$. You could double 5 and halve 6. This would give the same answer: $10 \times 3 = 30$.

3 Fill in the gaps.

	Problem	Double and halve	Product
e.g.	5×6	10×3	30
a	3×14	6×7	
b	5×18	10×9	
c	3×16		
d	5×22		
e	6×16		
f	4×18		

4 Here is a mental strategy for multiplying by 5.

	× 5	First multiply by 10	Then halve it	Multiplication fact
e.g.	14	140	70	$14 \times 5 = 70$
a	16			
b	18			
c	24			
d	32			
e	48			

5 Use your choice of strategy to find the product. Be ready to explain how you got the answer.

a 18×10 _____

b 14×100 _____

c $2.5 \text{ m} \times 10$ _____

d 34×10 _____

e 14×20 _____

f 150×5 _____

g 13×8 _____

h 9×40 _____

i $\$1.75 \times 10$ _____

j 8×60 _____

OXFORD UNIVERSITY PRESS

Extended practice

1 Use the split strategy to multiply by 15.

	× 15	× 10	Halve it to find × 5	Add the two answers	Multiplication fact
e.g.	12	120	60	120 + 60 = 180	12 × 15 = 180
a	16				
b	14				
c	20				
d	30				
e	25				

2 At the beginning of the year, Dee's mum gave her two spending-money choices.

- Choice 1: "Would you like $10 a week this year?"

- Choice 2: "Would you prefer 10c for the first four weeks, then double it for the next four weeks, then double it for the next four weeks, and so on for the rest of the year?"

- Dee remembered the ten trick and said, "52 weeks × $10 is $520. I'll take Choice 1 thanks, Mum."

Was this the better choice? How much would Dee have got if she'd taken Choice 2?

> Working-out space

3 Tran is reading books for his school's read-a-thon. He writes down how many pages he reads each day for a week.

Monday	48
Tuesday	48
Wednesday	48
Thursday	48
Friday	48
Saturday	45
Sunday	45

a Use mental strategies to find the total number of pages Tran reads during the week.

_____

b Explain the way you found the answer.

_____

Area model

You can work out multiplication problems by breaking the numbers down by place value and marking them off on grid paper.

This is called an area model, because as you calculate the total number of squares marked off, you are finding the area of the rectangle.

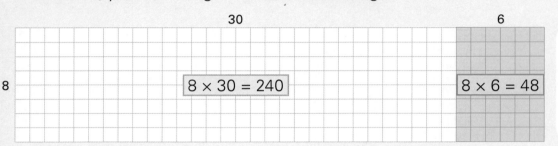

$$36 \times 8$$
$$8 \times 36 = 8 \times 30 + 8 \times 6$$
$$= 240 + 48$$
$$= 288$$

Guided practice

1 $7 \times 34 = 7 \times$ _____ $+ 7 \times$ _____

$=$ _____ $+$ _____

$=$ _____

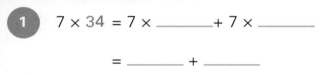

> Would the product be the same if I multiplied by the ones first?

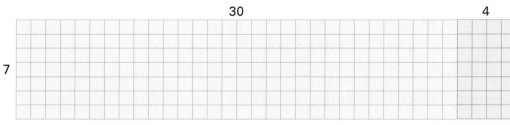

$7 \times 30 =$ ☐ $7 \times 4 =$ ☐

---

2 $5 \times 28 = 5 \times$ _____ $+ 5 \times$ _____

$=$ _____ $+$ _____

$=$ _____

$5 \times 20 =$ ☐ $5 \times 8 =$ ☐

Independent practice

Shade the model and fill in the blanks to find the product.

1 6 × 32

= _____ × _____ + _____ × _____

= _____ + _____

= _____

2 5 × 35

= _____ × _____ + _____ × _____

= _____ + _____

= _____

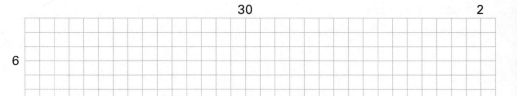

3 7 × 48

= _____ × _____ + _____ × _____

= _____ + _____

= _____

Contracted (short) multiplication

42 × 4 is the same as 2 × 4 and 4 tens × 4, so the answer is 8 plus 16 tens (160). You can make written multiplication short by writing a contracted algorithm. You start with the ones and then multiply each column in turn to find the product.

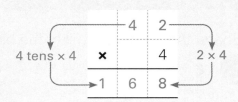

If you need to trade, you can do it like this.

3 tens × 4

4 × 4

3 tens × 4 = 12 tens
There is another 1 ten.
That makes 13 tens.

4 × 4 = 1 ten and 6 ones
1 ten goes in the tens column.

1 Complete the algorithms. Trade if necessary.

a
```
  1
  4  3
×    4
────────
```

b
```
  6  5
×    3
────────
```

c
```
  2  9
×    2
────────
```

d
```
  9  2
×    7
────────
```

e
```
  3  8
×    4
────────
```

2 Solve these problems in the same way.

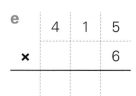

It works the same with larger numbers. Start at the ones, and complete each column in turn.

a
```
  1  2  5
×       2
──────────
```

b
```
  1  4  2
×       4
──────────
```

c
```
  2  5  3
×       3
──────────
```

d
```
  3  2  5
×       3
──────────
```

e
```
  4  1  5
×       6
──────────
```

f
```
  3  4  8
×       2
──────────
```

g
```
  4  7  5
×       3
──────────
```

h
```
  1  6  2  3
×          4
────────────
```

i
```
  1  2  7  2
×          5
────────────
```

j
```
  2  1  7  3
×          4
────────────
```

k
```
  1  2  3  2
×          8
────────────
```

28

OXFORD UNIVERSITY PRESS

Independent practice

1 Once you understand the short form of multiplication, it doesn't matter how big the number is that you are multiplying. Start at the ones column and complete each column in turn.

a

1	6	2	3
			4

b

1	7	3	4
			4

c

2	5	1	6
			3

d

4	2	3	0
			5

e

1	2	3	2	6
				3

f

2	1	5	3	8
				2

g

3	3	6	4	0
				7

h

2	3	8	5	2
				5

i

3	6	3	7	4
				5

j

2	4	7	3	7
				9

2 Find the product. Look for a pattern in the answers.

a

3	7	0	3	7
				3

b

3	7	0	3	7
				6

c

3	7	0	3	7
				9

d

7	4	0	7	4
				6

e

7	9	3	6	5
				7

f

7	4	0	7	4
				9

g

2	5	9	2	5	9
					3

h

1	2	6	9	8	4
					7

i

1	4	2	8	5	7
					7

3 To multiply an amount of money, start with the column of least value. Complete these in the same way as the example.

e.g.

$	1	¹2	•	²3	5
					4
$	4	9	•	9	0

a

$	2	7	•	2	5
					3
$			•		

b

$	1	8	•	7	5
					5
$			•		

c

$	1	4	•	5	5
					6
$			•		

The ten trick

When you are multiplying by a multiple
of ten, you need to remember the
"ten trick".

In the ten trick, everything
moves over one place.

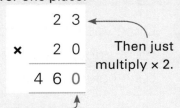

Then just
multiply × 2.

Put a ZERO as a space-filler. ⎯⎯⎯

4 Use the ten trick to complete these algorithms.

a			b			c			d			e		
	1	7		1	4		1	6		1	6		2	7
×	2	0	×	2	0	×	3	0	×	4	0	×	3	0
		0			0									

Multiplying by a 2-digit number
is like doing two multiplications
in one. You split the number
you are multiplying by.

What is 17 × 23?
There are two multiplications.

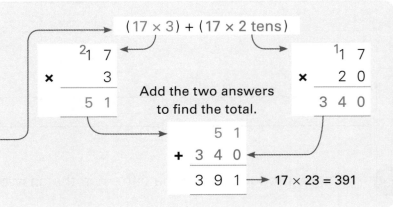

$(17 \times 3) + (17 \times 2 \text{ tens})$

Add the two answers
to find the total.

17 × 23 = 391

5 Split the numbers to multiply. Use separate paper to work out the answers to
questions d, e and f.

a What is 15 × 24?

Make two multiplications.

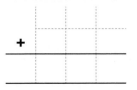

Add the answers.

15 × 24 = _____

b What is 16 × 23?

Make two multiplications.

Add the answers.

16 × 23 = _____

c What is 19 × 25?

Make two multiplications.

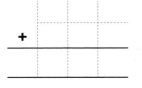

Add the answers.

19 × 25 = _____

d What is 16 × 39?

Make two multiplications.

e What is 15 × 37?

Make two multiplications.

f What is 19 × 45?

Make two multiplications.

OXFORD UNIVERSITY PRESS

Extended practice

An aeroplane flies millions of kilometres in its lifetime. This table shows the distances from Melbourne airport, in Australia, to other airports around the world.

Distance from Melbourne to:	
Adelaide, Australia	651 km
Bangkok, Thailand	7363 km
Chicago, USA	11 559 km
Darwin, Australia	3143 km
Edmonton, Canada	13 993 km
Frankfurt, Germany	16 308 km
Glasgow, Scotland	16 962 km
Honolulu, USA	8870 km
Istanbul, Turkey	14 619 km
Johannesburg, South Africa	10 326 km
Kuala Lumpur, Malaysia	6360 km
Los Angeles, USA	12 764 km

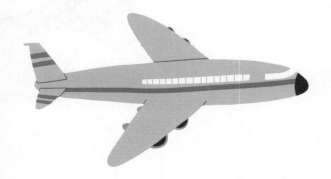

1 Choose an efficient multiplication method to find the answers. You may need extra paper for your calculations.

a   What is the distance of a return trip to Istanbul?

b   How far does a plane fly if it makes three return trips to Bangkok?

c   If a plane flies to and from Chicago eight times, how far does it fly?

d   A plane flies from Melbourne to Darwin and back twice a day for two weeks. What distance does it cover?

e   If a plane travelled to and from Johannesburg 50 times, would it have flown a million kilometres?

2 People can earn points for the distances they travel on certain airlines. ABC Airlines offers one point for every kilometre that its passengers fly.

a   Olivia flies from Melbourne to Adelaide on business each day and back home again from Monday to Friday. How many points does she earn in two weeks?

b   Tran travels from Kuala Lumpur to Melbourne once a month to visit his family. How many points does he earn in a year?

c   How many points does a family of four people earn by going on a holiday to Frankfurt?

3 A plane flies from Melbourne to Adelaide and back twice a day. How many kilometres does it fly in one week?

> 1 is a factor of every whole number.

A **factor** is a number that will divide evenly into another number:
2 is a factor of 4.

A **multiple** is the result of multiplying a number by a whole number:
6 is a multiple of 3 (3 × 2 = 6).

Guided practice

1 Circle the factors of each number.

e.g. **The factors of 10 are:**   ① ② 3 4 ⑤ 6 7 8 9 ⑩

a	The factors of 8 are:	1	2	3	4	5	6	7	8	
b	The factors of 5 are:	1	2	3	4	5				
c	The factors of 9 are:	1	2	3	4	5	6	7	8	9
d	The factors of 6 are:	1	2	3	4	5	6			
e	The factors of 2 are:	1	2							
f	The factors of 4 are:	1	2	3	4					
g	The factors of 7 are:	1	2	3	4	5	6	7		
h	The factors of 3 are:	1	2	3						

2 Write the first ten multiples of each number.

e.g. **10: 10, 20, 30, 40, 50, 60, 70, 80, 90, 100**

a 3: _____

b 6: _____

c 9: _____

d 2: _____

e 4: _____

f 8: _____

g 7: _____

h 5: _____

OXFORD UNIVERSITY PRESS

Independent practice

1 Write the factors of each number.

a   15 ☐☐☐☐   b   16 ☐☐☐☐☐

c   20 ☐☐☐☐☐☐   d   13 ☐☐

e   14 ☐☐☐☐   f   18 ☐☐☐☐☐☐

2 Which numbers between 21 and 30 have exactly:

a   two factors? _____   b   four factors? _____

c   three factors? _____   d   six factors? _____

3 a   List all eight factors of 24. _____

b   Which number between 30 and 40 has even more factors than 24? _____

List its factors. _____

4 The number 2 is a factor of every even number. Factors that are the same for more than one number are called *common factors*.

The factors of 16 are:	1	2	4	8	16	
The factors of 20 are:	1	2	4	5	10	20
The common factors of 10 and 20 are:	1	2	4			

a   The factors of 4 are: _____

The factors of 8 are: _____

The common factors of 4 and 8 are:

_____

b   The factors of 6 are: _____

The factors of 8 are: _____

The common factors of 6 and 8 are:

_____

c   The factors of 14 are: _____

The factors of 21 are: _____

The common factors of 14 and 21

are: _____

d   The factors of 12 are: _____

The factors of 18 are: _____

The common factors of 12 and 18

are: _____

5 For each row, circle the numbers that are multiples of the red number.

e.g.	3	3	6	9	13	15	23	27	30	34	39	40	42
a	5	15	21	25	40	50	57	60	65	69	75	85	100
b	4	8	12	22	24	26	28	30	34	36	40	42	48
c	8	8	12	16	20	24	30	32	36	44	48	56	60
d	7	14	20	21	27	28	35	37	42	47	49	56	60
e	9	9	12	18	21	24	27	36	39	45	55	63	72

6 How do you know that:

a   74 is a multiple of 2? _____

b   48 is a multiple of 3? _____

c   1001 is **not** a multiple of 10? _____

d   5551 is a **not** a multiple of 5? _____

7 When numbers share the same multiples, we call them **common** multiples.
List the multiples of 2 and 3 as far as 30. Circle the common multiples.

Multiples of 2:   2   4   6 _____

Multiples of 3:   3   6 _____

8 Find a common multiple of 4 and 5 between 1 and 30. _____

9 Find a common multiple of 2 and 3 between 31 and 40. _____

10 What is the **lowest** common multiple of:

a   6 and 9?   [ ]   b   3 and 4?   [ ]

c   5 and 7?   [ ]   d   3 and 5?   [ ]

e   5 and 9?   [ ]   f   4 and 7?   [ ]

OXFORD UNIVERSITY PRESS

Extended practice

1 In a biscuit factory, they make a lot of the same item. When the biscuits are ready, they have to decide how many should be put in each packet.

a If 50 biscuits were baked in an hour, they could put all 50 in one box. Find another five factors of 50 that will show the other ways that 50 biscuits could be packed. _____

b What would be a sensible number of biscuits to put in each packet? _____

2 A donut machine makes a batch of four donuts every minute.

a Circle the numbers of donuts that it is possible for the machine to make:

| 16 | 24 | 30 | 36 | 50 | 52 | 90 | 96 |

b How many donuts does the machine make in one hour?

c If the machine slowed down to three donuts a minute, which numbers in question 2a would it be possible for the machine to make? _____

d Which of the numbers of donuts can be made from both the faster and slower speeds? _____

3 Pencils at the Pixie Pencil Company come off the conveyor belt in batches of 96. Find all the options for the number of pencils that could go in a packet.

_____

4 The Bigfoot Sock Company makes 100 socks a day. Every sock is the same size and colour.

a How many different ways could the socks be packed?

_____

b Bigfoot's customers will not accept an odd number of socks. What are the options for the number of socks that could be put in a pack from one batch?

_____

2 is one of your factors.

2 is NOT one of your factors.

A factor is a whole number that will divide equally into another whole number. 1 is a factor of every whole number and 2 is a factor of half of all the whole numbers.

Every even number is divisible by 2.

Guided practice

1 Circle all the numbers that are exactly divisible by 2.

18   43   29   78   514   707   1000   2001   1234   990   2223   118

2 4 is an even number.

a Is every even number divisible by 4? _____

b Test your answer by circling the even numbers that are exactly divisible by 4.

2   4   6   8   10   12   14   16   18   20

22   24   26   28   30   32   34   36   38   40

3 All three of these numbers are exactly divisible by 4: 320, 716 and 5812. What do you notice about the divisibility of the red part (tens and ones) of each number?

_____

4 Circle all the numbers that are exactly divisible by 4.

112   620   425   426   428   340   342   716   714   410   412

5 Without using numbers that appear on this page, write:

a a 3-digit number that is exactly divisible by 2.

b a 3-digit number that is exactly divisible by 4.

c a 4-digit number that is exactly divisible by 2.

d a 4-digit number that is exactly divisible by 4.

OXFORD UNIVERSITY PRESS

There is a way to test for divisibility.

Test to see if a number can be divided exactly by:	It can if ...	Example
2	the number is even	In 135 792 the last digit is an even number so 135 972 is an even number. (135 792 ÷ 2 = 67 896)
3	the sum of the digits in the number is divisible by 3	In 24 the sum of the digits is 2 + 4 = 6. (6 ÷ 3 = 2)
4	the last two digits can be divided by 4	In 132, the last 2 digits are 32. 32 can be divided by 4 (132 ÷ 4 = 34)
5	the number ends in 5 or a 0	95 ends in 5 (95 ÷ 5 = 19).
6	the number is **even** and it is divisible by 3	78 is even and the sum of its digits is 7 + 8 = 15. 15 is divisible by 3 (78 ÷ 6 = 13).
8	the last three digits are a number that can be divided by 8	In 1048, the last 3 digits are 048. 48 is divisible by 8 (1048 ÷ 8 = 131).
9	the sum of the digits in the number is divisible by 9	In 153, the sum of the digits is 1 + 5 + 3 = 9. 153 is divisible by 9 (153 ÷ 9 = 17).
10	the number ends in a zero	543 210 ends in a zero so it is divisible by 10. (543 210 ÷ 10 = 54 321)

1   Use the divisibility tester for this activity. Circle the numbers that are exactly divisible by:

a (3)   411   207   433   513

b (5)   552   775   630   751

c (6)   711   702   522   603

d (8)   888   248   244   884

e (9)   819   693   539   252

f (10)   802   820   990   1001

2 A **prime** number has just two factors: 1 and itself. 37 is a prime number because it can only be divided by 1 and 37.

Use the divisibility tester to help you circle the only other prime number on this part of a 100 chart.

31	32	33	34	35	36	37	38	39	40

3 Numbers that have more than two factors are called **composite** numbers. 35 is a composite number. It has four factors: 1, 35, 5 and 7.

Which numbers in question 2 are exactly divisible by:

a   2 and 4?   _____   b   3 and 6?   _____

c   4 and 8?   _____   d   2 and 3?   _____

e   2, 4 and 8?   _____   f   3 and 11?   _____

g   2, 3, 4, 6 and 9?   _____

4 39 is divisible by 3. What are its other factors? _____

5 How do you know that 531 is divisible by 3? Circle one answer.

a   It has a 3 in it.   b   It is an odd number.   c   The sum of the digits is divisible by 3.

6 Circle the number that is divisible by 4.   4446   9324   2442   1234

7 Jack has 246 model cars. He wants to put them in groups of 4.

a   How could you explain to Jack that it is not possible to do that?

_____

b   Would it be possible to put them in groups of 3? _____

c   Explain your answer to question 7b.

_____

d   How many more cars would Jack need to be able to make groups of 4? _____

OXFORD UNIVERSITY PRESS

Extended practice

1 When a number is divisible by another number, it is also divisible by the factors of that number. For example, 2 and 3 are factors of 6. So, if a number is divisible by 6, it can also be divided by 2 and by 3. Prove it for yourself with the following.

Circle the numbers that are divisible by 6, 2 and 3.

24   54   72   96   48   78

2 Use the Venn diagram to show which of the following numbers are divisible by 3, by 4 and by both 3 and 4.

15   20   44   45
48   72   76   81
92   96

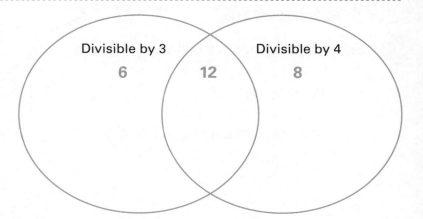

Divisible by 3

6

12

Divisible by 4

8

3 **a** How do you know that 306 is divisible by 6?

_____

b Find all the single-digit numbers that will divide exactly into 306.

_____

4 There is a number between 700 and 730 that is divisible by every single-digit number except 7. What is the special number? _____

5 Complete the Venn diagram to show some numbers that are divisible by 4, 5 and by both 4 and 5.

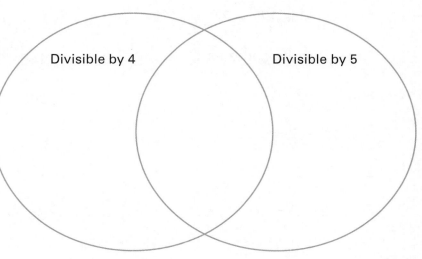

Divisible by 4

Divisible by 5

One written way to
solve a division problem
is to split the number you
are dividing by to find the quotient.

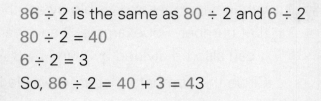

Let's share 86 marbles between us two.

86 ÷ 2 is the same as 80 ÷ 2 and 6 ÷ 2

80 ÷ 2 = 40

6 ÷ 2 = 3

So, 86 ÷ 2 = 40 + 3 = 43

Guided practice

1 Split these numbers to find the quotient.

a   What is 68 ÷ 2?

68 ÷ 2 is the same as 60 ÷ 2 and 8 ÷ 2

60 ÷ 2 = _____

8 ÷ 2 = _____

So, 68 ÷ 2 = _____ + _____ = _____

b   What is 69 ÷ 3?

69 ÷ 3 is the same as 60 ÷ 3 and 9 ÷ 3

60 ÷ 3 = _____

9 ÷ 3 = _____

So, 69 ÷ 3 = _____ + _____ = _____

c   What is 84 ÷ 2?

84 ÷ 2 is the same as 80 ÷ 2 and 4 ÷ 2

_____ ÷ 2 = _____

_____ ÷ 2 = _____

So, 84 ÷ 2 = _____ + _____ = _____

d   What is 124 ÷ 4?

124 is the same as 100 ÷ 4 and 24 ÷ 4

100 ÷ 4 = _____

24 ÷ 4 = _____

So, 124 ÷ 4 = _____ + _____ = _____

e   What is 122 ÷ 2?

122 ÷ 2 is the same as _____ ÷ 2

and 22 ÷ 2

_____ ÷ 2 = _____

22 ÷ 2 = _____

So, 122 ÷ 2 = _____ + _____ = _____

f   What is 145 ÷ 5?

145 ÷ 5 is the same as _____ ÷ 5

and _____ ÷ 5

_____ ÷ 5 = _____

_____ ÷ 5 = _____

So, 145 ÷ 5 = _____ + _____ = _____

Independent practice

Division can be set out in an algorithm. You put the number in a "box" and split it up. This is called short division. Imagine the problem is 42 ÷ 3. This is how it works:

Step 1
4 tens split into groups of three makes 1 group of three tens and 1 ten left over.

$$3\overline{)4^{1}2}$$ with quotient $1\,4$

Step 2
Trade the ten for 10 ones. Now there are 12 ones.

1 Find the quotient using the short division method.

a $4\overline{)5\,{}^{1}6}$ with quotient 1

b $2\overline{)3\,6}$

c $5\overline{)8\,5}$

d $6\overline{)7\,8}$

e $3\overline{)7\,2}$

f $7\overline{)8\,4}$

g $4\overline{)7\,{}^{3}6}$

h $5\overline{)9\,5}$

i $6\overline{)8\,4}$

j $3\overline{)8\,7}$

k $8\overline{)9\,6}$

l $7\overline{)9\,1}$

---

2 These problems contain larger numbers but you can solve them in the same way.

a $4\overline{)4\,6\,{}^{2}8}$ with quotient $1\,1$

b $5\overline{)5\,6\,0}$

c $3\overline{)6\,5\,1}$

d $2\overline{)8\,5\,0}$

e $6\overline{)6\,9\,6}$

f $3\overline{)9\,5\,4}$

g $5\overline{)5\,8\,5}$

h $7\overline{)7\,9\,8}$

i $2\overline{)6\,7\,4}$

j $6\overline{)6\,9\,0}$

k $3\overline{)6\,4\,5}$

l $4\overline{)8\,9\,6}$

m $3\overline{)3\,7\,8}$

n $7\overline{)7\,9\,1}$

o $2\overline{)8\,9\,8}$

p $6\overline{)6\,8\,4}$

When the digit in the first column cannot be divided, this is what you do.

1 There aren't enough hundreds to make groups of 2. We start with 11 tens.

11 tens split into groups of 2 = 5r1

$$2\overline{)\ \mathbf{1}\,1\,8}\qquad \overset{5}{}$$

2 Trade the ten for 10 ones. That makes 18 ones.

18 split into groups of 2 = 9

$$2\overline{)\ 1\,1^1 8}\qquad \overset{5\ 9}{}$$

3 Find the quotient.

a $2\overline{)\ 1\ 7^1 4}$ $\overset{8}{}$ b $3\overline{)\ 1\ 6\ 2}$ $\overset{5}{}$ c $3\overline{)\ 1\ 4\ 4}$ d $4\overline{)\ 1\ 3\ 6}$

e $6\overline{)\ 1\ 3\ 2}$ f $4\overline{)\ 2\ 6\ 8}$ g $5\overline{)\ 2\ 7\ 0}$ h $7\overline{)\ 3\ 9\ 9}$

i $9\overline{)\ 4\ 6\ 8}$ j $6\overline{)\ 2\ 8\ 2}$ k $8\overline{)\ 6\ 8\ 0}$ l $4\overline{)\ 3\ 7\ 2}$

Remember to write the digits in the correct columns.

m $3\overline{)\ 2\ 9\ 4}$ n $5\overline{)\ 3\ 9\ 5}$ o $7\overline{)\ 6\ 4\ 4}$ p $2\overline{)\ 1\ 9\ 8}$

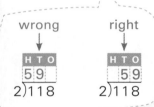

wrong | right
H T O | H T O
5 9 | 5 9
2)118 | 2)118

Sometimes the number you are dividing will not split equally. When this happens you have a remainder. This can be shown using "r" for remainder.
For example, 13 ÷ 3 = 4 r1.

4 Find the quotient. Use "r" to show the remainder.

a $4\overline{)\ 5\ 7}$ $\overset{\ \ \ r}{}$ b $2\overline{)\ 5\ 1}$ c $5\overline{)\ 7\ 7}$ d $6\overline{)\ 8\ 0}$

e $6\overline{)\ 6\ 9\ 3}$ f $3\overline{)\ 9\ 5\ 2}$ g $5\overline{)\ 5\ 8\ 2}$ h $7\overline{)\ 7\ 8\ 2}$

i $3\overline{)\ 1\ 6\ 7}$ j $6\overline{)\ 2\ 7\ 5}$ k $4\overline{)\ 2\ 6\ 5}$ l $5\overline{)\ 2\ 7\ 7}$

m $7\overline{)\ 2\ 9\ 3}$ n $9\overline{)\ 3\ 9\ 4}$ o $8\overline{)\ 5\ 4\ 7}$ p $2\overline{)\ 1\ 9\ 9}$

OXFORD UNIVERSITY PRESS

Extended practice

1 Not every number can be divided equally by other numbers. Write algorithms to find the quotient for these. Use remainders where necessary in the answers.

a 97 ÷ 5	**b** 72 ÷ 3	**c** 145 ÷ 6	**d** 386 ÷ 7

2 In real life, we have to work out what to do with remainders. You know that 7 ÷ 2 = 3 r1, 9 ÷ 2 = 4 r1 and 13 ÷ 2 = 6 r1. What is the best way to express the answers in these real-life situations?

a Seven donuts shared between two people.

b Two people are given nine marbles. How many can each person have?

c Two sisters share $13. How much do they each get?

3 At a school, there are 161 children in the six senior classes.

a To find the mean (average) number of students per class, divide the total number of students by the number of classes. The mean is:

b Complete the table to show the actual number that could be in each class. Two classes have been filled in. None of the other classes has the same number of students as any other class.

Class	Number of students
3W	25
3/4D	26
4M	
5S	
5/6H	
6T	

4 Three people share a prize of $100.

a Calculate how much money each should receive.

b They ask a bank to change the money so that they can each have a fair share. List the coins and notes that each might have.

5 At a chicken farm, 3000 eggs a day are packaged. They are put into boxes. Each box can hold 8 dozen eggs. How many boxes are needed for 3000 eggs?

What does a fraction look like?

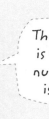

The number on the top is the numerator. The number on the bottom is the denominator.

A fraction can be part of a whole thing.

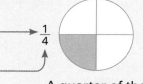

$\frac{1}{4}$

A quarter of the circle is blue.

A fraction can be part of a group of things.

$\frac{1}{2}$

Half of the beads are red.

Guided practice

1 Write the fractions in words and numbers.

a What fraction is red?

one sixth

$\frac{1}{}$

b What fraction is green?

one _____

$\frac{1}{}$

c What fraction is blue?

_____

$\frac{}{}$

d What fraction is black?

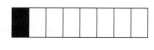

_____

$\frac{}{}$

2 Shade the shapes.

a Shade $\frac{3}{8}$

b Shade $\frac{3}{4}$

c Shade $\frac{2}{3}$

d Shade $\frac{3}{5}$

e Shade $\frac{5}{6}$

3 What fraction of the group is shaded?

a

b

c

d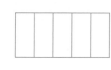

4 Shade each group to match the fraction.

a $\frac{3}{10}$

b $\frac{5}{12}$

c $\frac{2}{5}$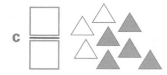

d $\frac{2}{4}$

OXFORD UNIVERSITY PRESS

Independent practice

1 Write the missing fractions on the number lines.

a

b

c

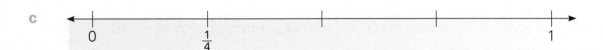

d

e

f

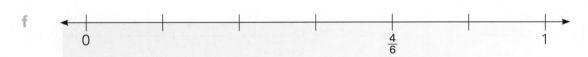

g

2 Which fraction in each pair is closer to 1? Use the number lines in question 1 to help.

a $\frac{1}{4}$ or $\frac{3}{8}$? _____

b $\frac{1}{3}$ or $\frac{1}{6}$? _____

c $\frac{1}{4}$ or $\frac{1}{8}$? _____

d $\frac{1}{5}$ or $\frac{1}{10}$? _____

e $\frac{1}{2}$ or $\frac{1}{3}$? _____

f $\frac{3}{4}$ or $\frac{7}{8}$? _____

g $\frac{7}{10}$ or $\frac{4}{5}$? _____

h $\frac{5}{8}$ or $\frac{1}{2}$? _____

i $\frac{7}{8}$ or $\frac{7}{10}$? _____

3 Which fractions are the same distance along the number line as $\frac{1}{2}$?

_____

4 Use the number lines on page 49 to help you order each group from **smallest** to **largest**.

a $\frac{4}{5}, \frac{1}{5}, 1, \frac{3}{5}, \frac{2}{5}$ _____

b $\frac{7}{10}, \frac{3}{10}, 1, \frac{9}{10}, \frac{2}{10}, \frac{6}{10}$ _____

c $\frac{1}{2}, \frac{1}{4}, \frac{1}{8}, \frac{1}{10}, \frac{1}{5}$ _____

d $\frac{3}{8}, \frac{3}{10}, \frac{3}{4}, \frac{3}{6}, \frac{3}{3}$ _____

e $\frac{2}{5}, \frac{2}{8}, \frac{2}{3}, \frac{2}{10}, \frac{2}{6}$ _____

5 Use the symbols > (is bigger than), < (is smaller than) or = to complete these number sentences.

a $\frac{3}{4}$ ☐ $\frac{7}{8}$ b $\frac{1}{4}$ ☐ $\frac{1}{8}$ c $\frac{3}{6}$ ☐ $\frac{1}{2}$ d $\frac{2}{3}$ ☐ $\frac{2}{6}$

e $\frac{3}{8}$ ☐ $\frac{1}{2}$ f $\frac{2}{4}$ ☐ $\frac{5}{8}$ g $\frac{9}{10}$ ☐ $\frac{4}{5}$ h $\frac{3}{5}$ ☐ $\frac{6}{10}$

i $\frac{5}{6}$ ☐ $\frac{2}{3}$

6

a Circle the two fractions that describe the position of the triangle on the number line.

$\frac{6}{8}$ and $\frac{6}{10}$ $\frac{6}{8}$ and $\frac{1}{4}$ $\frac{6}{8}$ and $\frac{3}{4}$

b Circle the fraction that describes how far from 1 the triangle is.

$\frac{2}{3}$ $\frac{2}{8}$ $\frac{2}{4}$

c Draw a diamond that is $\frac{3}{8}$ of the way from 0 to 1.

7 a Divide the rectangle into eighths.

b Shade $\frac{2}{8}$.

c What other fraction describes the fraction that you have shaded? _____

Extended practice

1　**a**　Divide the rectangle into three equal parts.

　　b　Shade one part.

　　c　Write two fractions that describe the shaded part.　_____

2　Write these fractions at the correct place on the number line.

a $\frac{1}{4}$　　　**b** $\frac{1}{2}$　　　**c** $\frac{3}{4}$　　　**d** $\frac{3}{8}$　　　**e** $\frac{7}{8}$

0　　　　　　　　　　　　　　　　　　　　1

3　Write a fraction that is:

　　a　bigger than a quarter but smaller than a half.

　　b　smaller than two-thirds but bigger than a half.

　　c　bigger than a third but smaller than a half.

　　d　bigger than five-sixths but smaller than one.

　　e　smaller than an eighth but larger than a twelfth.

4　Some people say it is impossible to fold a square of paper in half, then half again, and so on, more than eight times.

Is it true? How many times can you keep folding a piece of paper in half?

When you can go no further, write down the number of folds, then open the paper out and find the fraction that the folds have split the paper into.

- Number of folds: _____

- Fraction: _____

Was the result as you expected? Write a sentence to say how easy or how difficult you found this task.

_____

Adding and subtracting fractions

Working with fractions is like working with numbers when you first started school.

A fraction such as $\frac{3}{4}$ tells you the name of the fraction (denominator) and the number of parts that you have (numerator).

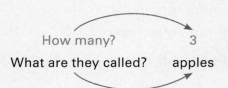

How many?   3
What are they called?   apples

numerator $\frac{3}{4}$ number
denominator   name

How many?   3
What are they called?   quarters

 $\frac{3}{4}$

Guided practice

It works in the same way for subtraction.

You can add fractions with the same denominator just as you do with ordinary objects.

1 apple + 2 apples = 3 apples

1 quarter + 2 quarters = 3 quarters

$\frac{1}{4}$ + $\frac{2}{4}$ = $\frac{3}{4}$

1 Fill in the gaps.

a 1 quarter + 1 quarter = ☐ quarters

$\frac{1}{4}$ + $\frac{1}{4}$ = $\frac{☐}{4}$

b 1 eighth + 2 eighths = ☐ eighths

$\frac{☐}{8}$ + $\frac{☐}{8}$ = $\frac{☐}{8}$

c 2 fifths + 2 fifths = ☐ fifths

$\frac{☐}{☐}$ + $\frac{☐}{☐}$ = $\frac{☐}{☐}$

d 2 sixths + 3 sixths = ☐ sixths

$\frac{☐}{☐}$ + $\frac{☐}{☐}$ = $\frac{☐}{☐}$

e

$\frac{3}{4}$ - $\frac{1}{4}$ = $\frac{☐}{☐}$

f

$\frac{☐}{☐}$ - $\frac{☐}{☐}$ = $\frac{☐}{☐}$

OXFORD UNIVERSITY PRESS

Independent practice

1 Write the number sentence.

e.g.

$$\frac{1}{4} \quad + \quad \frac{2}{4} \quad = \quad \frac{3}{4}$$

a

_____

b

_____

c

_____

d

_____

e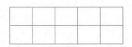

_____

2 Use two colours to shade each diagram to match the number sentence.

e.g. $\frac{2}{4} + \frac{1}{4} = \frac{3}{4}$    **a** $\frac{1}{5} + \frac{3}{5} =$ ☐

b $\frac{2}{6} + \frac{2}{6} =$ ☐   **c** $\frac{3}{8} + \frac{4}{8} =$ ☐

d $\frac{1}{3} + \frac{2}{3} =$ ☐   **e** $\frac{2}{10} + \frac{5}{10} =$ ☐

3 Use the diagrams to help complete the subtraction sentences.

e.g. $\frac{3}{4} - \frac{2}{4} = \frac{1}{4}$ 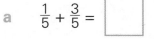   **a** $\frac{5}{8} - \frac{2}{8} =$ ☐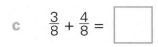

b $\frac{9}{10} - \frac{3}{10} =$ ☐ 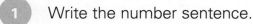   **c** $\frac{6}{6} - \frac{5}{6} =$ ☐

d $\frac{3}{5} - \frac{1}{5} =$ ☐    **e** $\frac{2}{3} - \frac{1}{3} =$ ☐

If the total comes to more than one whole, you use an **improper fraction** ($\frac{5}{4}$) or a **mixed number** ($1\frac{1}{4}$).

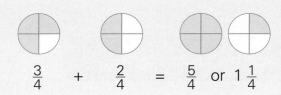

$\frac{3}{4}$ + $\frac{2}{4}$ = $\frac{5}{4}$ or $1\frac{1}{4}$

4 Write the answer as an improper fraction and as a mixed number.

a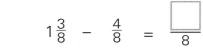

$\frac{5}{8}$ + $\frac{4}{8}$ = $\frac{\square}{8}$ or $1\frac{\square}{8}$

b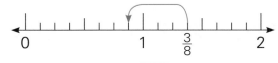

$\frac{4}{6}$ + $\frac{4}{6}$ = $\frac{\square}{6}$ or $1\frac{\square}{6}$

5 Use the number lines to help you add and subtract.

a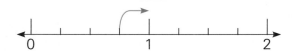

$\frac{3}{4}$ + $\frac{2}{4}$ = $\frac{\square}{4}$ = ___$\frac{\square}{\square}$

b

$1\frac{3}{8}$ − $\frac{4}{8}$ = $\frac{\square}{8}$

6 Complete the following. Use improper fractions and mixed numbers for the addition problems.

a $\frac{3}{4}$ + $\frac{3}{4}$ = $\frac{\square}{\square}$ or _____

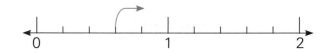

b $1\frac{5}{8}$ − $\frac{7}{8}$ = $\frac{\square}{\square}$

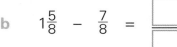

c $\frac{3}{5}$ + $\frac{4}{5}$ = $\frac{\square}{\square}$ or _____

d $1\frac{2}{6}$ − $\frac{4}{6}$ = $\frac{\square}{\square}$

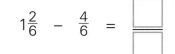

e $\frac{9}{10}$ + $\frac{4}{10}$ = $\frac{\square}{\square}$ or _____

f $1\frac{1}{3}$ − $\frac{2}{3}$ = $\frac{\square}{\square}$

OXFORD UNIVERSITY PRESS

Extended practice

It is possible to add different fractions, such as halves and quarters, but first you need to change them to the same type of fractions.

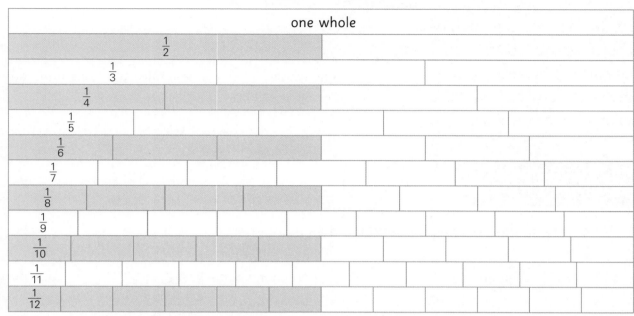

For example, what is $\frac{1}{2} + \frac{1}{4}$?

- On the fraction wall, you can see that $\frac{2}{4}$ is the same size as $\frac{1}{2}$.
- Change $\frac{1}{2}$ to $\frac{2}{4}$.
- Now you have $\frac{2}{4} + \frac{1}{4} = \frac{3}{4}$.

$$\frac{1}{4} + \frac{1}{2} = \frac{1}{4} + \frac{2}{4} = \frac{3}{4}$$

---

1 Use the fraction wall and the diagrams to help you complete these addition problems.

a

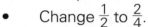

$$\frac{1}{6} + \frac{1}{2} = \frac{1}{6} + \frac{\square}{6} = \frac{\square}{6}$$

b

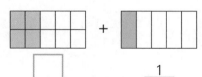

is the same as

$$\frac{\square}{10} + \frac{1}{\square} = \frac{\square}{10} + \frac{\square}{10} = \frac{\square}{10}$$

---

2 Use the fraction wall, diagrams or number lines to help solve these problems.

a $\frac{3}{10} + \frac{1}{5} = $ _____

b $\frac{2}{6} + \frac{1}{3} = $ _____

c $1\frac{1}{4} - \frac{1}{2} = $ _____

d $1\frac{1}{5} - \frac{3}{10} = $ _____

e $1\frac{1}{2} - 1\frac{1}{4} = $ _____

f $\frac{3}{8} + \frac{3}{4} = $ _____

g $\frac{1}{3} + \frac{1}{6} + \frac{1}{2} = $ _____

h $\frac{1}{4} + \frac{1}{2} + \frac{1}{8} = $ _____

Decimal fractions

If you split one whole into 10 equal parts, each part is a tenth. If you split one whole into 100 equal parts, each part is a hundredth. You can show tenths and hundredths as fractions and as decimals.

one whole
1

one-tenth
$\frac{1}{10}$
0.1

one-hundredth
$\frac{1}{100}$
0.01

Guided practice

1 Write the shaded part as a fraction and as a decimal.

e.g.

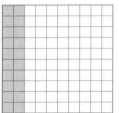

two-tenths
$\frac{2}{10}$
0.2

a

two-hundredths

_____

_____

b

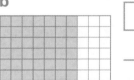

tenths

_____

c

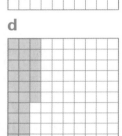

_____

_____

_____

d

_____

_____

_____

e

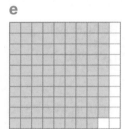

_____

_____

_____

2 Shade the diagrams to match the decimals.

| 0.4 | 0.04 | 0.15 | 0.7 | 0.99 |

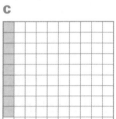 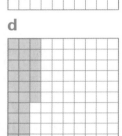 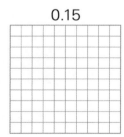 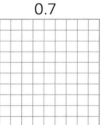 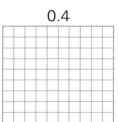

3 Write these as decimals.

a $\frac{3}{10}$ _____   **b** $\frac{23}{100}$ _____   **c** $\frac{3}{100}$ _____

4 Write these as fractions.

a 0.6 _____   **b** 0.77 _____   **c** 0.08 _____

A hundredth of a chocolate would not be very much, but there are even smaller fractions. If you split a hundredth into ten equal parts, each part is called **a thousandth**.

one-thousandth
$\frac{1}{1000}$
0.001

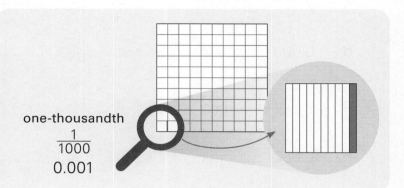

1 Fill in the gaps.

Four-thousandths of a piece of chocolate would be so small you'd need a magnifying glass to see it!

a 0._____

$\frac{\boxed{}}{1000}$

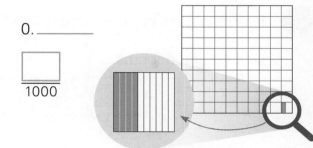

four-thousandths

b 0._____

$\frac{\boxed{}}{1000}$

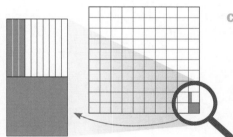

one-hundredth and three-thousandths

c 0._____

$\frac{\boxed{}}{1000}$

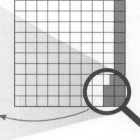

one tenth, two-hundredths and four-thousandths

2 Write these as decimals.

a $\frac{125}{1000}$ _____

b $\frac{8}{1000}$ _____

c $\frac{87}{1000}$ _____

d $\frac{2}{1000}$ _____

e $\frac{22}{1000}$ _____

f $\frac{99}{1000}$ _____

3 Write as a fraction:

a   0.005 _____

b   0.255 _____

c   0.101 _____

d   0.035 _____

e   0.999 _____

f   0.009 _____

4 Write this number using digits:

fourteen ones, six-tenths, two-hundredths and seven-thousandths _____

5. Complete these by writing the symbols > (is bigger than), < (is smaller than) or =.

a 0.01 _____ 0.001

b $\frac{3}{1000}$ _____ 0.003

c $\frac{25}{1000}$ _____ 0.25

d 0.003 _____ 0.2

e $\frac{125}{1000}$ _____ 0.125

f $\frac{6}{1000}$ _____ 0.01

g 0.02 _____ $\frac{2}{1000}$

h 1 _____ 0.999

i $\frac{19}{1000}$ _____ 0.19

j 0.052 _____ $\frac{52}{1000}$

k 0.430 _____ 0.043

l 0.999 _____ $\frac{999}{1000}$

6. Fill in the gaps on these decimal number lines.

a Count in tenths.

0   [ ] [ ] [ ] [ ]   0.5   [ ] [ ] [ ] [ ]   0.1

b Count in hundredths.

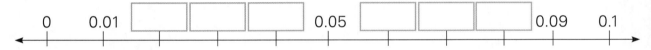

0   0.01   [ ] [ ] [ ]   0.05   [ ] [ ] [ ]   0.09   0.1

c Count in thousandths.

0   [ ] [ ] [ ] [ ]   0.005 [ ]   0.007 [ ] [ ]   0.01

7. Use the number lines to help you order these from **smallest** to **largest**.

a 0.2   0.5   0.1   0.9   0.4

_____

b 0.04   0.07   0.02   0.06   0.03

_____

c 0.007 0.004 0.008 0.002 0.001

_____

d 0.2   0.3   0.02   0.002   0.1

_____

e 0.1   0.11   0.2   0.22   0.15

_____

f 0.5   0.05   0.005   0.555   0.055

_____

OXFORD UNIVERSITY PRESS

1 Write the position of the triangle on each number line.

a _____

0.06   0.07

b _____

0.04   0.05

2 There are 100c in $1. So, one cent is $\frac{1}{100}$ of a dollar. It can also be written as $0.01.

How can five cents be written with a dollar sign and a decimal? _____

3 Write the following with a dollar sign and a decimal:

a 25c _____   b 8c _____   c $\frac{15}{100}$ of a dollar _____

d 75c _____   e 20c _____   f $\frac{80}{100}$ of a dollar _____

g 115c _____   h 2 dollars and $\frac{2}{10}$ of a dollar _____

4 To work out $2.90 × 3 on a calculator, you could press seven keys like this:

| 2 | . | 9 | 0 | × | 3 | = |

Show how it could be done by pressing just six keys:

| | | | | | = |

5 Some cafés show their prices using just one decimal place.

Using the normal way of writing money, find the cost of:

a A small coffee and a large muffin. [ ]

b A large coffee and two fruit scones. [ ]

c A small and a large coffee, a small muffin and two plain scones. [ ]

d A large coffee and **one** plain scone. [ ]

e Two large coffees, **one** plain scone and two fruit scones. [ ]

Menu

Coffee
Small: $3.2
Large: $3.9

Muffins
Small: $2.4
Large: $4.7

Scones (2 per serve)
Plain: $3.7
Fruit: $4.2

The symbol % stands for *per cent*. It means *out of a hundred*. 1% means one out of a hundred. It can be written as a fraction, as a decimal or as a percentage.

The amount shaded is:

$\frac{1}{100}$ fraction

0.01   decimal

1%   percentage

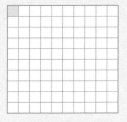

Another way of saying 100% is 1, or one whole.

Guided practice

1   Write each shaded part as a fraction, as a decimal and as a percentage.

a

Fraction $\frac{3}{100}$

Decimal _____

Percentage _____

b

Fraction $\frac{\quad}{\quad}$

Decimal _____

Percentage _____

c

Fraction $\frac{\quad}{\quad}$

Decimal _____

Percentage _____

d

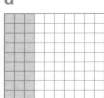

Fraction $\frac{\quad}{\quad}$

Decimal _____

Percentage _____

e

Fraction $\frac{\quad}{\quad}$

Decimal _____

Percentage _____

f

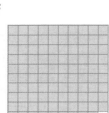

Fraction $\frac{\quad}{\quad}$

Decimal _____

Percentage _____

---

2   Shade the grid. Fill the gaps.

a

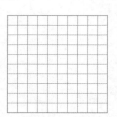

Fraction $\frac{20}{100}$

Decimal _____

Percentage _____

b

Fraction $\frac{\quad}{\quad}$

Decimal _____

Percentage   15%

c

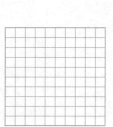

Fraction $\frac{\quad}{\quad}$

Decimal _____

Percentage   75%

d

Fraction $\frac{55}{100}$

Decimal _____

Percentage _____

OXFORD UNIVERSITY PRESS

1 Fill in the gaps to match the percentages, decimals and fractions.

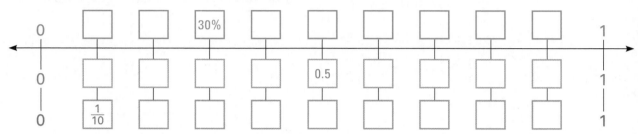

0 [ ] [ ] 30% [ ] [ ] [ ] [ ] [ ] [ ] 1

0 [ ] [ ] [ ] [ ] 0.5 [ ] [ ] [ ] [ ] 1

0 $\frac{1}{10}$ [ ] [ ] [ ] [ ] [ ] [ ] [ ] 1

---

2 Complete the table.

	Fraction	Decimal	Percentage
a	$\frac{5}{100}$		
b			25%
c		0.75	
d	$\frac{99}{100}$		
e	$\frac{9}{10}$		
f			40%
g		0.1	
h	$\frac{2}{100}$		
i		0.3	
j			100%
k	$\frac{1}{2}$		
l			1%

3 Write **true** or **false**.

a $10\% = \frac{1}{10}$ _____

b $0.01 < 1\%$ _____

c $0.2 = \frac{25}{100}$ _____

d $35\% = \frac{35}{100}$ _____

e $\frac{7}{10} < 75\%$ _____

f $0.9 > 9\%$ _____

g $\frac{2}{100} > 20\%$ _____

h $95\% = 0.95$ _____

i $100\% > 1$ _____

---

4 Compare the fractions and percentages.

a $\frac{1}{2}$ of this square is shaded.

Shade the same amount of this square.

$\frac{1}{2}$ is the same as [ ]%

b [ ] of this square is shaded.

Shade the same amount of this square.

[ ] is the same as [ ]%

c [ ] of this square is shaded.

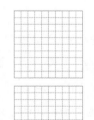

Shade the same amount of this square.

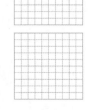

[ ] is the same as [ ]%

5 Order each row from **smallest** to **largest**.

a   0.03   20% $\frac{2}{100}$

_____

b   0.05   6%   0.5

_____

c   5% $\frac{1}{2}$ $\frac{55}{100}$

_____

d $\frac{1}{4}$ 40%   0.04

_____

e   70% $\frac{3}{4}$ 0.07

_____

f   10%   0.01 $\frac{11}{100}$

_____

6 Follow the instructions to colour these circles.

30% red, 0.4 blue, $\frac{30}{100}$ yellow

7 Write the fraction of triangles that are green as a decimal, as a fraction and as a percentage.

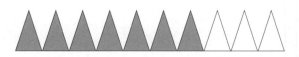

_____

8 Follow the instructions to colour these diamonds.

40% red, 20% blue, $\frac{30}{100}$ yellow, 5% green, 5% white

9 There are 20 beads on the string. Colour them these percentages: 50% red, 25% blue, 25% yellow.

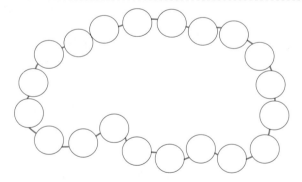

10 a   Choose a way to colour 25% of these beads. Leave the rest of the beads white.

b   Write the white part of the string of beads as a fraction, as a decimal and as a percentage.

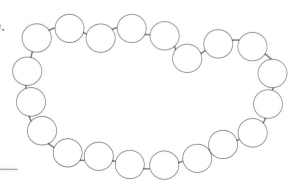

_____

Extended practice

1 If someone offers you 50% of their apple, it's the same as offering a half.
Complete the table to show what you would get if you were offered these items.

	Item	Percentage offered	Fraction	Number
a	Box of 20 donuts	50%		
b	Pack of 50 pencils	10%		
c	Tin of 80 cookies	25%		
d	Bag of 1000 marbles	1%		

2 In some situations it is possible to have more than 100%.

a If you cut a 1-metre piece of string and Sally said she needed a second piece that was 50% of the length of the first one, how long would it be?

b If Sally asked for another piece that was 100% of the length of the first one, how long would it be?

c If Sally asked for a fourth piece that was 200% of the length of the first one, how long would it be?

3 You need a knowledge of percentages to change the size of drawings in a computer program such as Microsoft Word. You will need a computer for the next activity.

a Open a new Word file.

b Choose a shape on **autoshapes** (PC) or **basic shapes** (Mac).

c Click and drag to draw a shape.

d Double-click on the shape so that you can format (change) it.

e Choose **Size**.

f Look for the **Scale** icon and change 100% to 120%. Then click **OK**.

g Take note of what happened. Experiment to see how you can double the size of the shape.

Write a short report about the way that entering different percentages can change a shape.

Year 5 want to raise money for an end-of-year party. They decide to buy fruit, cut it up and sell 100 fruit salads at a stall on "Fruit Salad Friday". They want to make a profit. This means that they sell the fruit for more than it costs to buy it.

FRUIT SALAD FRIDAY
EAT HEALTHY FOOD!
EAT CHEAP FOOD
FRUIT SALAD: $1.50

Guided practice

The less it costs to prepare the fruit salad, the more profit they will make.

1 Look at the sign. How much money will Year 5 take at the stall if they sell all 100 fruit salads? _____

2 If the fruit costs $150 to buy, Year 5 will not make any profit. How much profit will they make if the cost of the fruit is:

   a $100?   **b** $75?   **c** $50?   **d** $25?

   _____   _____   _____   _____

3 Year 5 decide to cut up five fruits into the fruit salads. How much would it cost if they bought:

   a 1 kg of each fruit? _____

   b 2 kg of each fruit? _____

   c 500 g of each fruit? _____

   d 5 kg of each fruit? _____

Grapes $10 kg
Oranges $3 kg
Bananas $2 kg
Pears $2.50 kg
Apples $4 kg

4 Flora's Fruit Shop offers a 10% discount if Year 5 buy 10 kg of each fruit.

   a What would be the total price before discount if Year 5 bought 10 kg of each fruit? _____

   b What would be the discount? _____

   c What would be the new price of the fruit? _____

5 If Year 5 bought 5 kg of each fruit, how much profit would they make? _____

Independent practice

1 Year 5 want to make a profit of at least $50, so they don't want to spend more than $100. If they buy 5 kg of each fruit, how much over their budget are they? _____

2 Year 5 need to spend less on the fruit. They decide to buy only 2.5 kg of grapes.

   a Circle any of the following that describe 2.5 kg of grapes compared to 5 kg apples:

      50%   a quarter   a half   0.5   0.75   25%

   b How much does 2.5 kg of grapes cost? _____

3 Flora's Fruit Shop send the fruit, along with an invoice to show how much Year 5 owe.

   a Write the cost for each type of fruit.

   b Write the total price of all the fruit.

   c Year 5 can get a 10% discount. Fill in the amount of the discount.

   d Write the new discounted total.

Flora's Fruits			
Description	**Quantity**	**Price per kg**	**Cost**
Apples	5 kg	$4.00	$20.00
Pears	5 kg	$1.50	
Oranges	5 kg	$3.00	
Bananas	5 kg	$2.00	
Grapes	2.5 kg	$10.00	
Total:			
10% discount if you pay by tomorrow. Discount:			
Discounted total:			

4 How much under their $100 budget will Year 5 be after buying the fruit? _____

5 The students need to buy 100 plastic spoons and **either** 100 plastic bowls **or** 100 plastic cups. Calculate the price for each option.

Working-out space

Cups $16.50 for 100

Bowls $22.00 for 100

Spoons $5.50 for 100

GST (Goods and Services Tax) is a tax that has to be paid for some purchases. A percentage of the cost is added to the price. The percentage can change.

Pete's Plastics

Item	Quantity	Unit price	Cost
Spoons	100	5c	$5.00
Cups	100	15c	$15.00
		Total price of goods	$20.00
		GST (10%)	
		Total:	

6 The class used spoons and cups. On Fruit Salad Friday, GST was 10%. Fill in the GST amount and total on the receipt.

7 Fill the gaps to show what the receipt would have looked like if Year 5 had bought 100 spoons and 100 bowls.

Pete's Plastics

Item	Quantity	Unit price	Cost
Spoons	100	5c	
Bowls	100	20c	
		Total price of goods	
		GST (10%)	
		Total:	

8 Two furniture shops are selling the same tables and chairs. One shows the price without GST. The other shows the price including GST.

Fill in the amounts to see which shop has the better price for a table and four chairs.

Chair $20 plus GST   Table $120 plus GST

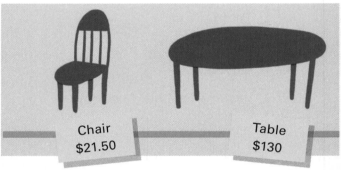

Chair $21.50   Table $130

Furniture World

Item	Quantity	Unit price	Cost
Table	1	$120.00	
Chairs	4	$20.00	
		Price of goods	
		GST (10%)	
		Total:	

Furniture For You

Item	Quantity	Unit price	Cost
Table	1	$130.00	
Chairs	4	$21.50	
		Total price of goods (including GST)	

9 Both shops have an end-of-year sale. They offer 10% off the final prices. What is the new price for a table and four chairs at each shop?

a   Furniture World: _____

b   Furniture For You: _____

OXFORD UNIVERSITY PRESS

Extended practice

1 A receipt for a restaurant meal shows a price of $90.20, including GST.
Circle the price of the meal **before** 10% GST was added.

$80 $82 $80.20 $82.20

---

2 If GST is 10%, the price before the tax was added is $\frac{10}{11}$ of the final price.
You can see that this is true by using a final price of $11 for a meal:

- $11.00 ÷ 11 = $1.00. $1.00 × 10 = $10.00 for the meal before GST.

- $10.00 + 10% GST = $11.00

If the meal costs $22.00, what is the price before GST? _____

> You will need access to a computer and a program such as Microsoft Excel for the next activity.

3 Not all amounts divide easily by 11. You can use a spreadsheet (such as Excel) to work out the amounts easily. Follow these steps in a new Excel workbook.

a In cells A1, B1 and C1 type:

	A	B	C	D
1	Full price	Before GST	GST amount	

b Click on cell B2 and then on the Formula Bar. If you don't see the Formula Bar, click on **View** and then on **Formula Bar**.

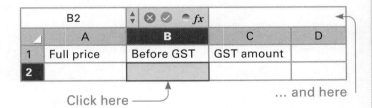

Click here   ... and here

c To tell the computer to find $\frac{10}{11}$ of the price, type in the Formula Bar: =A2/11*10. (This formula tells the computer to divide the amount in cell A2 by 11 and then multiply it by 10.)

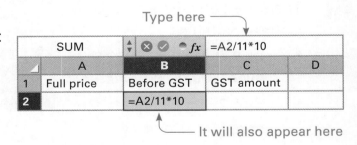

Type here

It will also appear here

d Press **Return**.

e Click on cell A2 and enter the full price as $99.99.

f Press **Return**, and watch the price before GST appear in cell B2.

---

4 Calculate the GST amount and enter it into the GST column on the spreadsheet.

We use number patterns every day. You probably learned your first number pattern before you started school.

1, 2, 3, 4, 5, 6, 7, 8, 9, 10. Coming, ready or not!

Guided practice

1 The rule for this number pattern is: *The numbers increase by two each time.* Continue the pattern.

Position	1	2	3	4	5	6	7	8	9
Number	1	3	5						

2 Find the rule, then continue each number pattern. Write a rule for each pattern using the words **increase** or **decrease**.

a

Position	1	2	3	4	5	6	7	8	9
Number	100	98	96	94					

Rule: _____

b

Position	1	2	3	4	5	6	7	8	9
Number	$\frac{1}{2}$	1	$1\frac{1}{2}$						

Rule: _____

3 There are two different rules in these patterns.

- Rule 1: If the number is **even**, you divide by 2.
- Rule 2: If the number is **odd**, you take away 1 and then you divide by 2.

Follow the two rules to complete the table.

Number	10	12	15
Is it even?	Yes ÷ 2		
Answer	5		
Is it even?	No – 1, ÷ 2		
Answer	2		
Is it even?	Yes ÷ 2		
Answer	1		
Is it even?	No – 1, ÷ 2		
Answer	0	0	

4 It takes four steps to get to zero for the starting numbers in question 3. How many steps does it take to get to zero if the starting number is:

a 8? _____

b 25? _____

OXFORD UNIVERSITY PRESS

Independent practice

1 Read the rule to complete each table.

a Start at 5 and increase by 4 each time.

Term	1	2	3	4	5	6	7	8	9	10
Number	5									

b Start at 10. Decrease by 0.5 each time.

Term	1	2	3	4	5	6	7	8	9	10
Number	10	9.5								

2 Continue these patterns for the first ten terms. Write a rule for each pattern.

a 0, 0.2, 0.4, 0.6, _____ Rule: _____

b $\frac{3}{4}$, $1\frac{1}{2}$, $2\frac{1}{4}$, 3, _____ Rule: _____

3 The rules for question 3 on page 68 can be shown in a diagram.

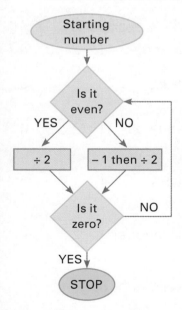

Using 18 as a starting number, the steps that follow the rules are:

$18 \div 2 = 9$
$(9 - 1) \div 2 = 4$
$4 \div 2 = 2$
$2 \div 2 = 1$
$(1 - 1) \div 2 = 0$

Write the steps that take 22 to zero.

_____

_____

4 This diagram shows new rules. Follow the rules to take these numbers to zero.

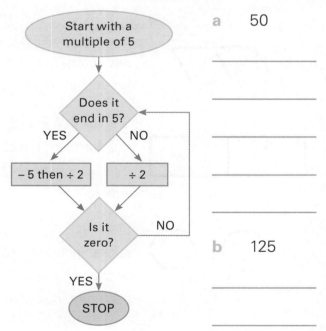

a 50

_____

_____

_____

_____

_____

b 125

_____

_____

_____

_____

_____

5 Number patterns can help in creating shape patterns. These patterns are made with sticks. Fill in the gaps.

Pattern of sticks	Rule for making the pattern	How many sticks are needed?				
e.g.	Start with 3 sticks. Increase the number of sticks by 3 for each new triangle.	Number of triangles	1	2	3	4
		Number of sticks	3	6	9	12
a	Start with 4 sticks. Increase the number of sticks by ____ for each new diamond.	Number of diamonds	1	2	3	4
		Number of sticks				
b	Start with ____ sticks. Increase the number of sticks by ____ for each new hexagon.	Number of hexagons	1	2	3	4
		Number of sticks				
c	Start with ____ sticks. Increase the number of sticks by ____ for each new pentagon.	Number of pentagons	1	2	3	4
		Number of sticks				

6 These stick patterns are made in a different way. Complete the rule and write the number of sticks for each term.

Pattern of sticks	Rule for making the pattern	How many sticks are needed?				
e.g.	Start with 3 sticks. Increase the number of sticks by 2 for each new triangle.	Number of triangles	1	2	3	4
		Number of sticks	3	5	7	9
a	Start with 4 sticks. Increase the number of sticks by ____ for each new square.	Number of squares	1	2	3	4
		Number of sticks	4			
b	Start with ____ sticks. Increase the number of sticks by ____ for each new hexagon.	Number of hexagons	1	2	3	4
		Number of sticks	6			

7 How many sticks would be needed at the 10th term for the squares and hexagons in question 6?

Squares: _____

Hexagons: _____

OXFORD UNIVERSITY PRESS

Extended practice

1 Imagine that you work for an advertising company. Your boss wants you to deliver advertising leaflets in a town with 1000 houses. She knows that some houses will have a No Junk Mail sign. This table gives information about whether the houses are likely to accept junk mail.

Number of houses	1	2	3	4	5	6	7	8	9	10	11	12	13	14	15
Junk mail OK?	yes	yes	yes	yes	no	yes	yes	yes	yes	no	yes	yes	yes	yes	no

a Circle the ending for the rule that describes the pattern. The number of houses that do not accept junk mail is:

5 out of 5   1 out of 5   4 out of 5   1 out of 4   1 out of 2

b How many out of 10 houses do not want junk mail?

c How many out of 100 houses would not want junk mail?

d How many out of 1000 houses would not want junk mail?

e How many leaflets will you need to deliver?

2 A toy company is ordering wheels for toy cars. Each car has four wheels.

This table shows how many wheels are needed for the cars. Complete the table for the first ten terms of the pattern.

Number of cars										
Number of wheels										

3 The company does not make the same number of cars every week.
How many wheels would be needed for:

a 25 cars?   **b** 100 cars?   **c** 250 cars?   **d** 1000 cars?

4 The toy company decides it needs to have extra wheels in case some get lost. They decide to get an extra wheel for every 25th car. Re-calculate the number of wheels that need to be ordered for:

a 50 cars   **b** 100 cars   **c** 350 cars   **d** 1250 cars

Number operations and properties

Working with number sentences is a bit like putting on your clothes. Sometimes the order of doing things does not matter—and sometimes it does!

Changing the order of putting on clothes ...

Left then right, or right then left ...	Sock then shoe ...	Shoe then sock...

Changing the order of numbers ...

Addition	Subtraction
3 + 2 = ? or 2 + 3 = ? ✔	3 − 2 = ? ✔ or 2 − 3 = ? ✘

Guided practice

1 Try changing the number order with each operation.

Addition

	Number sentence	Change the order	Same answer?
e.g.	3 + 2 = ?	2 + 3 = ?	Yes
a	14 + 2 = ?		
b	20 + 12 = ?		
c	15 + 10 = ?		

Subtraction

	Number sentence	Change the order	Same answer?
e.g.	3 − 2 = ?	2 − 3 = ?	No
a	14 − 2 = ?		
b	20 − 12 = ?		
c	15 − 10 = ?		

Multiplication

	Number sentence	Change the order	Same answer?
e.g.	3 × 2 = ?	2 × 3 = ?	Yes
a	14 × 2 = ?		
b	20 × 12 = ?		
c	15 × 10 = ?		

Division

	Number sentence	Change the order	Same answer?
e.g.	3 ÷ 2 = ?	2 ÷ 3 = ?	No
a	14 ÷ 2 = ?		
b	20 ÷ 12 = ?		
c	15 ÷ 10 = ?		

2 Complete these sentences.

Can you see how addition and multiplication are connected?

a The answer is the same if you change the order of the numbers for addition and _____.

b The answer is **not** the same if you change the order of the numbers for _____.

OXFORD UNIVERSITY PRESS

1 Changing the order of the numbers can help with mental calculations. Put these into an order that will help you to solve the problems easily.

e.g. **17 + 18 + 3 = ?** Change to **17 + 3 + 18 = 38 (17 + 3 = 20, then add 18)**

a 15 + 17 + 5 = ? _____

b 23 + 19 + 7 = ? _____

c 5 × 14 × 2 = ? _____

d 4 × 13 × 25 = ? _____

2 If there are three numbers in a subtraction problem, does it matter which number you subtract first? Find out with these number sentences.

a 25 – 10 – 5 = _____, 25 – 5 – 10 = _____

b 36 – 12 – 6 = _____, 36 – 6 – 12 = _____

c 28 – 15 – 8 = _____, 28 – 8 – 15 = _____

3 If there are three numbers in a division problem, does it matter which number you divide by first? Find out with these number sentences.

a 16 ÷ 2 ÷ 4 = _____, 16 ÷ 4 ÷ 2 = _____

b 36 ÷ 6 ÷ 2 = _____, 36 ÷ 2 ÷ 6 = _____

c 72 ÷ 2 ÷ 9 = _____, 72 ÷ 9 ÷ 2 = _____

4 Addition and subtraction are connected. Multiplication and division are connected. Show how one "undoes" the other by completing these tables.

	Addition and subtraction		Multiplication and division	
	Addition sentence	Subtraction sentence	Multiplication sentence	Division sentence
e.g.	17 + 8 = 25	25 – 8 = 17	3 × 5 = 15	15 ÷ 5 = 3
a	14 + 12 =		9 × 8 =	
b	35 + 15 =		25 × 4 =	
c	22 + 18 =		15 × 10 =	
d	19 + 11 =		20 × 6 =	

An equation is a number sentence in parts.
The parts balance each other.

$$2 \times 3 \quad = \quad 4 + 2$$

5 Complete these equations.

a
$$4 \times 2 \quad = \quad \boxed{} + 6$$

b
$$\boxed{} \div 2 \quad = \quad 3 + 6$$

c
$$16 \div 2 \quad = \quad 2 \times \boxed{}$$

d
$$\boxed{} - 14 \quad = \quad 3 + 7$$

e
$$40 \div 2 \quad = \quad 4 \times \boxed{}$$

f
$$9 \times 2 \quad = \quad \boxed{} \div 2$$

g
$$2 \times 7 \quad = \quad \boxed{} + 6$$

h
$$\boxed{} - 20 \quad = \quad 5 \times 6$$

i
$$30 \div 3 \quad = 100 \div \boxed{}$$

You can use equations to make calculation simpler.

6 Which of the following would balance $60 \div 2 \div 5$?

$$2 \div 60 \div 5 \qquad 5 \div 2 \div 60 \qquad 5 \div 60 \div 2 \qquad 60 \div 5 \div 2$$

7 Which of the following would **not** balance $17 + 19$?

$$2 \times 3 \times 6 \qquad 12 + 2 + 12 \qquad 56 - 20 \qquad 360 \div 10$$

8 Which of these is **not** correct?

$$4 \times 15 = 15 \times 4 \qquad 4 + 15 = 15 + 4 \qquad 15 + 4 = 4 + 15 \qquad 15 \div 4 = 4 \div 15$$

9 Find three different ways to balance the first part of the equation.

e.g.	$2 \times 3 \times 5$	$60 \div 2$	$5 + 15 + 10$	$100 - 50 - 20$
a	$5 + 20 + 8$			
b	$50 \div 2$			
c	$72 - 25$			
d	$6 \times 2 \times 10$			
e	$3 + 23 + 12$			
f	$40 \div 5 \div 2$			

OXFORD UNIVERSITY PRESS

Extended practice

So that we solve mathematic problems properly, we use this order of operations:

Brackets first.
Division and multiplication second.
Addition and subtraction last.

1 Write the answers to these pairs of number sentences.

Look for the problem in each pair that is easier to solve.

	Problem 1		Problem 2	
a	14 − 13 + 7 =		14 + 7 − 13 =	
b	49 − 24 + 25 =		25 − 24 + 49 =	
c	35 − 10 + 25 =		35 + 25 − 10 =	
d	175 − 50 + 25 =		175 + 25 − 50 =	

2 These pairs of number sentences look similar, but the answers are different.

	Problem 1		Problem 2	
a	7 + 2 × 3 =		(7 + 2) × 3 =	
b	10 − 8 ÷ 2 =		(10 − 8) ÷ 2 =	
c	15 ÷ 3 + 2 =		15 ÷ (3 + 2) =	
d	10 × 5 + 15 =		10 × (5 + 15) =	

3 Explain why the answer to 4 + 3 × 5 is different to the answer to (4 + 3) × 5.

------

4 When you read a word problem, things need to be done in the right order so that you arrive at the correct answer. Here is an example:

Tran had ten $1 coins. He lost four coins at playtime (so he had $6). His mother felt sorry for him and doubled the amount he had left ($6 × 2). How much did he have? ($12)

a To solve the problem we could write a number sentence. However, doing the following calculation will not give the right answer: 10 − 4 × 2. Why?

b Write a number sentence that would solve the problem correctly.

------

5 Write a story to suit this number sentence: (12 + 6) ÷ 3

When you are measuring, it is important to be as accurate as possible. The length of this pencil is **not** 8 cm.

Guided practice

1 The pencil above is more than 8 cm.
Circle the best estimate for its actual length: 9 cm   10 cm   11 cm   12 cm

2 Write the length of each red line above it.

e.g. 6 cm

a

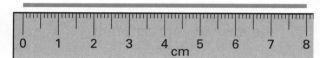

b

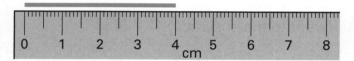

c

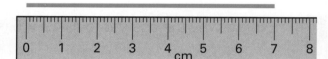

3 Write the length of the red lines in centimetres and millimetres, and in centimetres with a decimal.

e.g. 5 cm 2 mm or 5.2 cm   **a**   7 cm 1 mm or

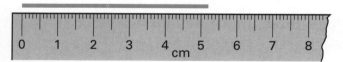

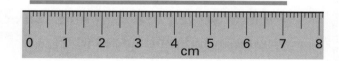

b

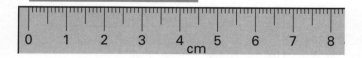

c

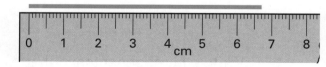

4 Use a ruler to measure these lines. Write the lengths as you did in question 3.

a _____

b _____

c _____

OXFORD UNIVERSITY PRESS

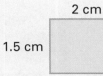
There is a quicker way to find a perimeter.

The total length of all the sides of the rectangle is 2 cm + 1.5 cm + 2 cm + 1.5 cm = 7 cm, so the perimeter is 7 cm.

2 cm

1.5 cm   1.5 cm

2 cm

1 You could find the perimeter of the rectangle above by adding 2 cm and 1.5 cm (= 3.5 cm) and then doubling to get the answer (= 7 cm). Explain why.

_____

2 Calculate the perimeters of these rectangles without using a ruler. (They are not drawn to scale.)

a _____   b _____   c _____

6 cm   4 cm   5 cm

3 cm   2 cm   3 cm

3

a   How many lines would you need to measure to find the perimeter of this square?

b   What is the perimeter of the square?

4 Find the perimeter of each 2D shape. How many lines did you need to measure?

a   b   c   d

| Perimeter | | Perimeter | | Perimeter | | Perimeter |
| Number of lines | | Number of lines | | Number of lines | | Number of lines |

In this topic, you have used centimetres and millimetres as units of length.
Metres and kilometres can also be used.

- 10 mm = 1 cm
- 100 cm = 1 m
- 1000 m = **1 km**

Use the information to complete these length conversion tables.

5

×10 cm mm ÷10		
e.g.	**1 cm**	**10 mm**
a	2 cm	
b	7 cm	
c		90 mm
d	3.5 cm	
e		75 mm

6

×100 m cm ÷100		
e.g.	**1 m**	**100 cm**
a		200 cm
b	3 m	
c	7 m	
d		500 cm
e	$9\frac{1}{2}$ m	

7

×1000 km m ÷1000		
e.g.	**1 km**	**1000 m**
a	2 km	
b		4000 m
c		55000 m
d	9.5 km	
e	8.5 km	

You can use different units of length to measure the same object.
For example, a pencil could be described as 9 cm long or 90 mm long.

8 Which two units of length would you use for these?

a The length of a pencil sharpener

_____ _____

b The height of a door

_____ _____

c The length of an eraser

_____ _____

d The length of a road

_____ _____

9 Find the perimeters of these shapes. Write the answer in millimetres, and in centimetres with a decimal.

a

b

c

d

Perimeter:

_____ mm

_____ cm

Perimeter:

_____ mm

_____ cm

Perimeter:

_____ mm

_____ cm

Perimeter:

_____ mm

_____ cm

OXFORD UNIVERSITY PRESS

Extended practice

1 There is a special right-angled triangle called a "3, 4, 5" triangle, because the sides are always in that proportion. A triangle with sides in those proportions always has a right angle in it. It doesn't matter if the unit of measurement is centimetres, metres, millimetres or even kilometres.

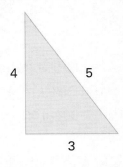

On a separate piece of paper, draw triangles whose sides are multiples of "3, 4, 5" numbers. You could start with a 3 cm by 4 cm by 5 cm triangle and then a "6, 8, 10" triangle. With your teacher's permission, you could draw a triangle measuring 3 m by 4 m by 5 m.

2 Draw a square that has a perimeter of exactly 10 cm.

3 Audrey has a pencil that is $14\frac{1}{2}$ cm long. Write the length in as many different ways as you can.

$14\frac{1}{2}$ cm

_____

4 **a** This line is 3.2 cm long. Increase the length by 12.3 cm.

_____

b Write the total length of the line in two different ways. _____ _____

5 Write the perimeters of these **regular** shapes in two ways. They are not drawn to scale.

a

2.1 cm

Perimeter:

_____ mm

_____ cm

b

3.3 cm

Perimeter:

_____ mm

_____ cm

c

1.9 cm

Perimeter:

_____ mm

_____ cm

d

2.4 cm

Perimeter:

_____ mm

_____ cm

e

3.5 cm

Perimeter:

_____ mm

_____ cm

Area is the surface of something, such as the area of the top of a table. In a classroom, we usually measure area in square centimetres (cm²) or square metres (m²). It can also be measured in square millimetres (mm²), hectares (ha) or square kilometres (km²).

Area is always measured in squares.

Guided practice

 The shapes have square centimetres drawn on them.
Write the areas.

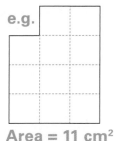

e.g.

Area = 11 cm²

a Area = _____ cm²

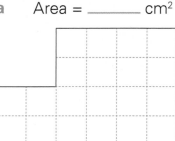

b Area = _____ cm²

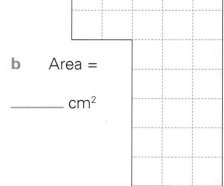

c Area = _____ cm²

d Area = _____ cm²

e Area = _____ cm²

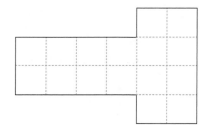

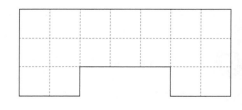

 Write the area of these rectangles.

e.g. Area = 6 cm²

a Area = _____ cm²

b Area = _____ cm²

c Area = _____ cm²

d Area = _____ cm²

e Area = _____ cm²

OXFORD UNIVERSITY PRESS

To find the area of a rectangle, you need to know:

- how many squares fit on a row

- how many rows there are.

2 rows
3 squares on a row
Area = **2** rows of **3** cm^2
Area = 6 cm^2

1 Find the areas of these rectangles.

a

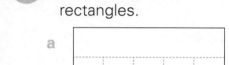

_____ rows _____ squares on a row

Area = _____ rows of _____ cm^2 Area = _____

b

_____ rows

_____ squares on a row

Area = _____ rows

of _____ cm^2

Area = _____

c

_____ rows

_____ squares on a row

Area = _____ rows

of _____ cm^2

Area = _____

d

Area = _____

e

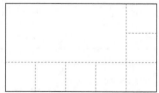

Area = _____

f

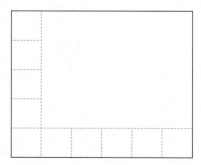

Area = _____

g

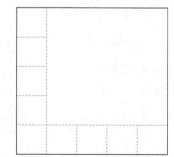

Area = _____

If you know the length and width of a rectangle, you can find the area by imagining how many squares will fit on a row and how many rows there are. You can see how it happens on this rectangle.

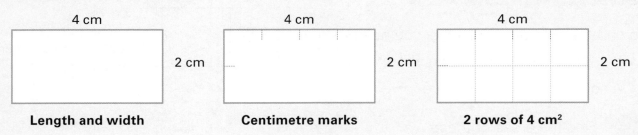

Length and width   **Centimetre marks**   **2 rows of 4 cm²**

2 Use a method of your choice to find the area of each rectangle.

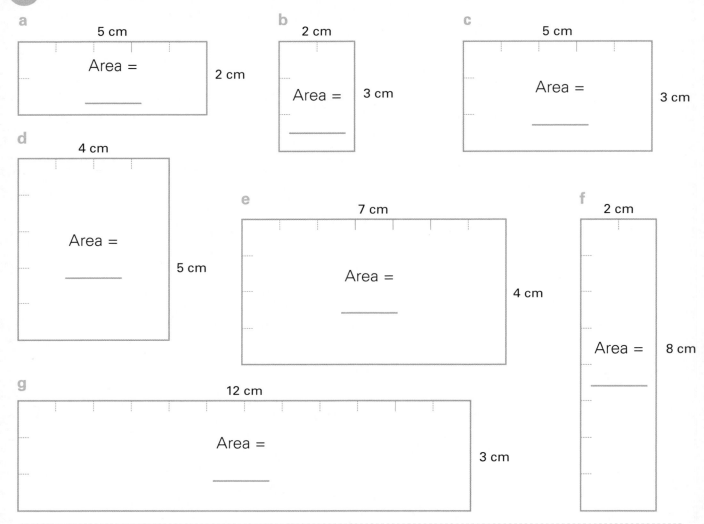

a
5 cm

Area =

_____

2 cm

b
2 cm

Area =

_____

3 cm

c
5 cm

Area =

_____

3 cm

d
4 cm

Area =

_____

5 cm

e
7 cm

Area =

_____

4 cm

f
2 cm

Area =

_____

8 cm

g
12 cm

Area =

_____

3 cm

3 Use the dimensions of each rectangle to help find its area. They are not drawn to actual size.

a
3 cm

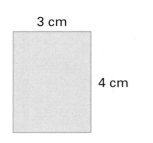

4 cm

Area = _____

b
8 cm

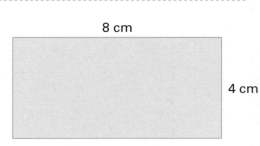

4 cm

Area = _____

OXFORD UNIVERSITY PRESS

Extended practice

1 Measure each rectangle to find its area.

a
Area = _____

b
Area = _____

c
Area = _____

---

2 If you can split a shape into rectangles, you can find its area. Find the area of each shape.

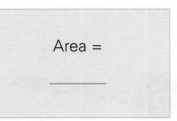

a
3 cm
2 cm
B
4 cm
2 cm
A
2 cm

Area of A = _____

Area of B = _____

Total area = _____

b
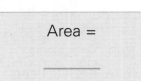
5 cm
A
B
3 cm
4 cm
2 cm

Area of A = _____

Area of B = _____

Total area = _____

c
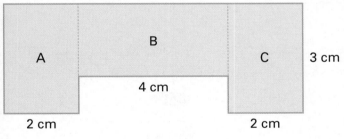
A
B
C
3 cm
4 cm
2 cm
2 cm

Area of A = _____   Area of C = _____

Area of B = _____   Total area = _____

d
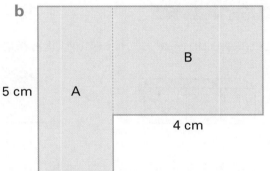
3 cm
A
2 cm
1 cm
B
2 cm
2 cm
C

Area of A = _____

Area of B = _____

Area of C = _____

Total area = _____

e

Total area = _____

f

Total area = _____

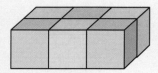

Volume is the space something takes up. It is measured in cubes. This centimetre cube model has a volume of 6 cubic centimetres (6 cm³).

Capacity is the amount that can be poured into something. We normally use litres (L) and millilitres (mL). This spoon has a capacity of 5 mL.

Everything that takes up space has volume— even me!

Guided practice

1 Write the volume of each centimetre cube model.

e.g.

Volume = 3 cm³

a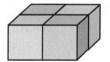

Volume = _____ cm³

b

Volume = _____ cm³

c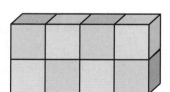

Volume = _____ cm³

d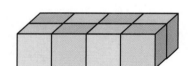

Volume = _____ cm³

e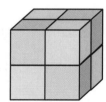

Volume = _____ cm³

2 A cup has a capacity of about 250 mL.

Circle the most likely capacity of the following containers.

a

6 mL 60 mL
600 mL 6 L

b

2 mL 20 mL
200 mL 2 L

c

30 mL 300 mL
3 L 30 L

d

80 mL 800 mL
8 L 80 L

3 What is something that has a capacity of about a litre? _____

OXFORD UNIVERSITY PRESS

Independent practice

1 Write the volume of each centimetre cube model.

a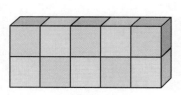

b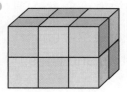

c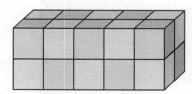

Volume = _____ cm³   Volume = _____ cm³   Volume = _____ cm³

d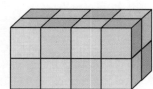

e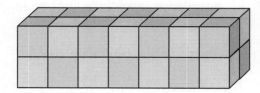

Volume = _____ cm³   Volume = _____ cm³

---

2 a How many centimetre cubes would be needed to make this model?   _____

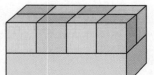

b What is the volume?   _____

c If there were three layers the same, what would the volume be?   _____

---

3 a How many centimetre cubes are on the bottom layer of this box?   _____

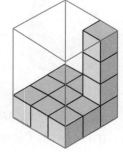

b How many layers does the box hold?   _____

c What is the volume of the box?   _____

---

4 How do you know that the volume of this model is 8 cm³?

1 cm

2 cm

4 cm

_____

---

5 Look at the model in question 4. What would the volume be if the height were:

a 2 cm? _____   b 3 cm? _____   c 4 cm? _____   d 5 cm? _____

6 The capacity of the milk carton could also be written as 1000 mL, because there are 1000 mL in a litre.

Complete the table to convert between millilitres and litres.

Milk 1 L

	Litres	Millilitres
e.g.	1 L	1000 mL
a		2000 mL
b		
c	3 L	
d		9000 mL
e	5.5 L	
f		2500 mL
g	1.25 L	
h		3750 mL

(×1000 Litres → Millilitres, ÷1000)

7 Order each row from smallest to largest.

a   2 L 400 mL, 2.5 L, 2350 mL   _____

b $\frac{1}{2}$ L, 450 mL, 0.35 L   _____

c   1850 mL, $1\frac{3}{4}$ L, 1.8 L   _____

d $\frac{1}{4}$ L, 200 mL, 20 mL   _____

8 Which of these drink containers holds closest to half a litre?   _____

A
Orange 750 mL

B
Fruit juice 200 mL

C
Water 375 mL

D
Apple 600 mL

9 Look at the containers in question 8. Use the information next to each jug to shade it to the correct level when the drinks have been poured in. Write the amounts in millilitres.

a 1 fruit juice and 1 apple drink

Amount:

_____ mL

2L
1L

b 2 orange drinks

Amount:

_____ mL

2L
1L

c 1 water and 1 apple drink

Amount:

_____ mL

2L
1L

OXFORD UNIVERSITY PRESS

Extended practice

1 **a** What is the volume of this rectangular prism? _____

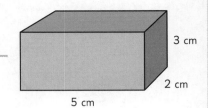

3 cm
2 cm
5 cm

b Explain why you can find the volume by multiplying the length by the width by the height.

_____

2 Calculate the volume of each rectangular prism.

a
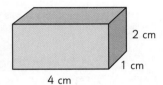
2 cm
1 cm
4 cm

b
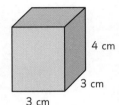
4 cm
3 cm
3 cm

c
10 cm
4 cm
4 cm

Volume: _____   Volume: _____   Volume: _____

d
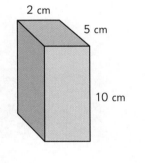
2 cm
5 cm
10 cm

e
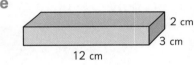
2 cm
3 cm
12 cm

f
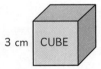
3 cm   CUBE

Volume: _____   Volume: _____   Volume: _____

3 Scientists have proved that 1 cm^3 takes up exactly the same space as 1 mL of water. This is hard to prove in real life. Try it for yourself.

You need 20 centimetre cubes and a measuring jug that goes up in 10-mL steps.

What to do:

- Put 30 mL of water in the measuring jug.
- Put 10 cubes in the water. What is the new level?
- Put 5 more cubes in the water. What is the new level?
- Put 5 more cubes in the water. What is the new level?
- Did it work like it was supposed to do? Write a few lines about what you did. If it didn't work, suggest a reason.

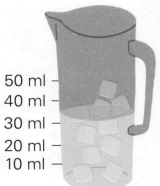

50 ml
40 ml
30 ml
20 ml
10 ml

Milligrams (mg)   Grams (g)   Kilograms (kg)   Tonnes (t)

Each unit of mass is 1000 times heavier than the one before it.

Mass tells us how heavy something is. We use four units of mass.

Guided practice

Dog   Apple   Train   Grains of sand

1 Under each picture, write the most likely unit of mass (milligrams, grams, kilograms or tonnes).

a _____   b _____   c _____   d _____

---

2 Complete the tables to convert between units of mass.

	Tonnes ×1000 Kilograms ÷1000	
e.g.	1 t	1000 kg
a	2 t	
b		4000 kg
c		1500 kg
d	3.5 t	
e	1.25 t	

	Kilograms ×1000 Grams ÷1000	
e.g.	1 kg	1000 g
a		2000 g
b	5 kg	
c	3.5 kg	
d		1250 g
e	0.5 kg	

	Grams ×1000 Milligrams ÷1000	
e.g.	1 g	1000 mg
a	5 g	
b		3000 mg
c	1.5 g	
d		2500 mg
e	0.5 g	

---

3 Write the mass of each box in grams.

a 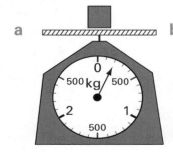   b 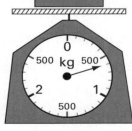   c   d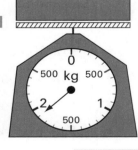

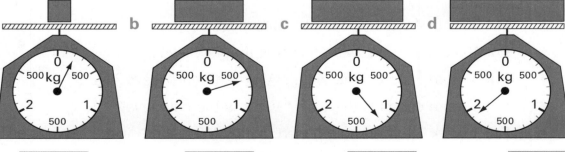

Mass: ☐   Mass: ☐   Mass: ☐   Mass: ☐

OXFORD UNIVERSITY PRESS

The mass of this box can be written as $2\frac{1}{2}$ kg, 2.5 kg or 2 kg 500g.

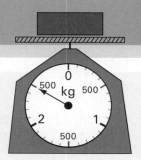

1 Complete this table.

	Kilograms and fraction	Kilograms and decimal	Kilograms and grams
e.g.	$2\frac{1}{2}$ kg	2.5 kg	2 kg 500 g
a		1.5 kg	1 kg 500 g
b	$2\frac{1}{4}$ kg		
c		4.75 kg	
d		1.3 kg	

2 Not all scales have the same increments (markings). Look carefully at these scales and write the masses in kilograms and grams, and in kilograms with a decimal.

a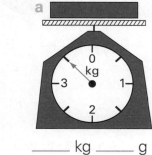

b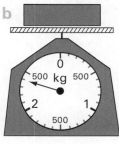

c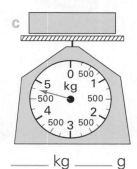

d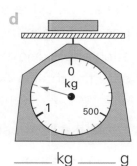

a _____ kg _____ g _____ kg

b _____ kg _____ g _____ kg

c _____ kg _____ g _____ kg

d _____ kg _____ g _____ kg

3 Look at the scales in question 2. Would you use scale A, B, C or D if you needed to have:

a 100 g of butter? ☐

b 650 g of flour? ☐

c 4.25 kg of potatoes? ☐

d 2.5 kg of apples? ☐

4 Draw pointers on the scales to show the mass of each box.

a 1 kg 500 g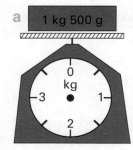

b 850 g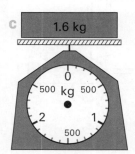

c 1.6 kg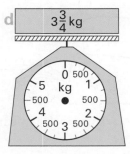

d $3\frac{3}{4}$ kg

5 **a** Reorder the trucks, from lightest at the front to heaviest at the back.

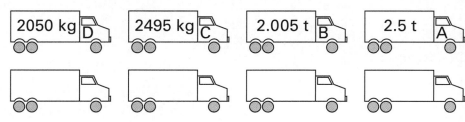

b Which trucks are carrying less than $2\frac{1}{2}$ t? _____

c Which two trucks are carrying a combined mass of 4.5 t? _____

d True or false? The combined mass of all the trucks is more than 9 t. _____

6 These four apples have a total mass of half a kilogram, but none weigh exactly the same. Write a possible mass for each apple. Make sure the total is $\frac{1}{2}$ kg.

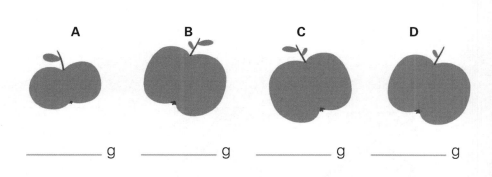

_____ g   _____ g   _____ g   _____ g

7 Circle the best estimate for the mass of these objects.

An elephant   A can of drink   A pencil sharpener   A Year 5 student

a   45 kg   **b**   3.5 g   **c**   1 g   **d**   35 kg
   450 kg   35 g   15 g   350 kg
   4500 kg   350 g   150 g   3500 kg

This truck can carry 2 tonnes.

8 Is the truck strong enough to carry the three boxes? _____

OXFORD UNIVERSITY PRESS

Extended practice

The mass of all four apples on page 90 was less than the mass of one record-breaking apple. Use the information in the table to complete the activities.

Record-breaking fruit or vegetable	Where and when?	Mass
Apple	Japan, 2005	1.849 kg
Cabbage	UK, 1999	57.61 kg
Lemon	Israel, 2003	5.265 kg
Peach	USA, 2002	725 g
Pumpkin	USA, 2009	782.45 kg
Strawberry	UK, 1983	231 g
Pear	Australia, 1999	2.1 kg
Blueberry	Poland, 2008	11.28 g

1 **a** Order the fruits and vegetables from lightest to heaviest.

_____

_____

b How much heavier than the cabbage is the pumpkin?

c How much heavier than the pear is the lemon?

d Which fruit is 1124 grams heavier than the heaviest peach? _____

e If strawberries like the heaviest one were sold in boxes of around a kilogram, how many would there be in a box?

f By how many grams is the heaviest strawberry heavier than the heaviest blueberry?

2 A group of seven Year 5 students found that they had a total mass of 273.854 kg.

a Divide the total mass by the number of students to find the average mass of a student in the group.

b Round the numbers and find out how many of the students it would take to balance the world's heaviest pumpkin.

3 Sol bought three apples. The first had a mass of 125 g. The second had a mass of 133 g and the mass of the third was 117 g. Use the same process as in question 2 to find the average mass of one apple. _____

Analogue clock   Digital clock

There are two main types of clock: analogue clocks and digital clocks. Analogue clocks have been around for hundreds of years. Digital clocks are more recent.

Times before noon are called am times. Times between noon and midnight are called pm times.

Guided practice

1 Write the times under the clocks. Use **am** and **pm**.

e.g. Waking up   **a**   At school   **b**   Doing homework   **c**   In bed

| 6:30 am |

d   Eating lunch   **e**   Getting dressed   **f**   Going home   **g**   Fast asleep

2 Some digital clocks have an indicator to show **am** and **pm**. Draw each time on the analogue clock. Write whether each time is **am** or **pm**.

   a   **b**   **c**   **d**

pm indicator off

pm indicator on

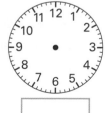  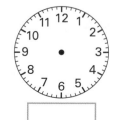 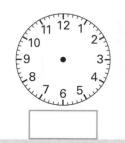

OXFORD UNIVERSITY PRESS

1 On a 24-hour clock, the times continue past 12 to 13, 14, and so on. 24-hour times are usually written as four digits with no spaces. Midnight is 0000. Fill in the 24-hour and am/pm times on this timeline.

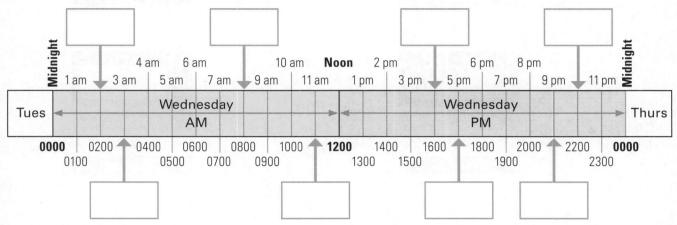

2 Convert these times to 24-hour times. For example, 8:15 am becomes 0815.

a	10 am	b	3:30 pm	c	2:20 pm	d	7:11 am

e	9:48 pm	f	7:11 pm	g	9:48 am	h	12:29 am

3 Write these events as 24-hour and am/pm times.

	Event	am/pm time	24-hour time
a	The time I leave for school		
b	The time I eat dinner		
c	The time I leave school		
d	The time I go to bed		

4 Owen's football match starts at 1420 and lasts for 45 minutes. Show the starting and finishing times on the analogue and digital clocks.

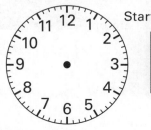 Starting time

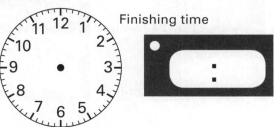 Finishing time

5 Fill in the gaps to show these times in four different ways.
Remember to use the pm indicator if necessary.

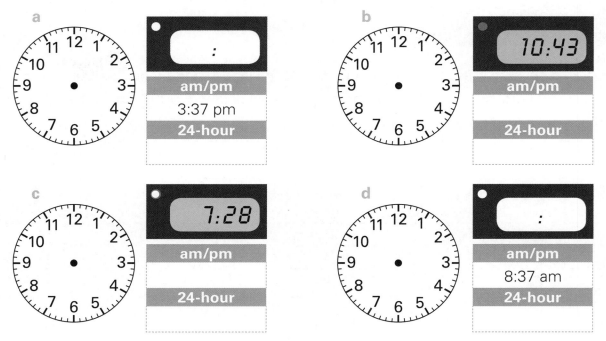

a

:

am/pm

3:37 pm

24-hour

b

10:43

am/pm

24-hour

c

7:28

am/pm

24-hour

d

:

am/pm

8:37 am

24-hour

---

6 This is Sam's timetable for Friday at school. Use the information to complete the activities.

Friday	
Sport	9:00
Maths	10:00
Recess	11:00
Reading groups	11:18
Journal writing	12:15
Lunch	13:00
Art & craft	14:00
Story	15:00

Literacy session

a At what time does the Mathematics lesson begin? (**Use am/pm time**.)

b When does the lunch break start? (**Use am/pm time**.)

c How long does Recess last?

d Lunchtime starts with 10 minutes "eating time". How much playtime does Sam have after that?

e How long does the Literacy session last?

f Estimate the time that Story reading begins. (**Use 24-hour time**.)

---

7 Digital clocks are used for 24-hour time as well as am/pm times.
Rewrite the times on these 24-hour clocks.

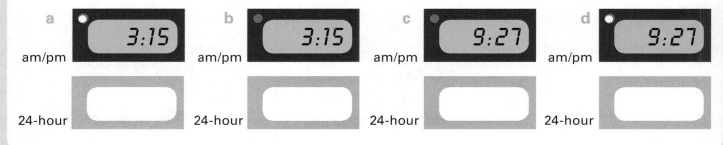

a

3:15

am/pm

24-hour

b

3:15

am/pm

24-hour

c

9:27

am/pm

24-hour

d

9:27

am/pm

24-hour

OXFORD UNIVERSITY PRESS

Extended practice

Puffing Billy

FROM BELGRAVE			
Belgrave	dep:	10:30	11:10
Menzies Creek	arr:	10:53	11:33
Menzies Creek	dep:	11:05	11:35
Emerald	dep:	11:20	11:53
Lakeside	arr:	11:30	12:08
Lakeside	dep:	...	12:20
Cockatoo	arr:	...	12:35
Gembrook	arr:	...	13:00

1 A train called Puffing Billy was built over 100 years ago. Puffing Billy got its name because it is a steam engine.

Above you can see part of a timetable for people who want to take a ride on Puffing Billy. Use the information to complete the following activities.

a How long does the 10:30 train take to get from Belgrave to Menzies Creek?

b How long does the 11:10 train wait at Menzies Creek?

c How much longer does the 10:30 train take to get from Belgrave to Lakeside?

d How long does the 11:10 train wait at Lakeside?

e How long does the journey take from Belgrave to Gembrook?

f Imagine there is a new summer service from Belgrave to Gembrook. The train leaves at 4:05 pm and takes the same length of time as the 11:10 train. At what 24-hour time will it arrive at Gembrook?

g The 11:10 train waits at Gembrook for an hour. It then returns to Belgrave, taking the same amount of time as the outward journey. At what 24-hour time does it arrive back at Belgrave?

A polygon is a closed shape with three or more straight sides. None of the sides cross over each other.

This polygon has **parallel** sides.
Parallel lines go in the same direction.

 This is **not** a polygon.

 This is a polygon.

A circle is a 2D shape, but it is not a polygon.

Guided practice

1 **a** Colour the polygons. Tick the polygons that have parallel sides.

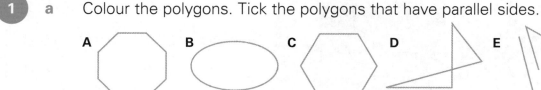

A   B   C   D   E   F

b Explain why the unshaded shapes are **not** polygons.

- **B** is not a polygon because _____.

- _____ is not a polygon because _____.

- _____ is not a polygon because _____.

2 Most polygons are named after the number of angles in the shape. Write each polygon's name. Use the word bank to help with spelling.

Word bank	triangle
pentagon	octagon
quadrilateral	hexagon

a    **b** 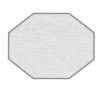   **c**   **d**   **e**

_____ _____ _____ _____ _____

3 A polygon is either **regular** or **irregular**. Regular shapes have all sides and angles the same. Irregular ones do not. Shade the regular polygons. Draw stripes on the irregular polygons. Add arrows on any pairs of parallel lines you see.

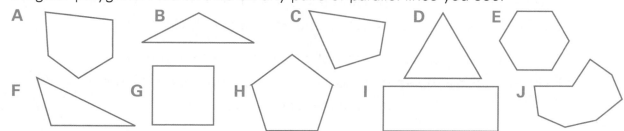

A   B   C   D   E

F   G   H   I   J

All triangles have three sides. Triangles can be named according to the lengths of their sides and the sizes of their angles.

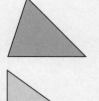 **Scalene triangle**: No sides are the same length. No angles are equal.

 Isosceles triangle: Two sides are the same length. Two angles are equal.

 Right-angled triangle: There is a right angle in the triangle.

 Equilateral triangle: All sides are the same length. All the angles are equal.

1 This rectangular pattern is made from triangles. Colour it according to the types of triangles:

- green for scalene triangles.
- yellow for right-angled triangles.
- blue for isosceles triangles.
- red for equilateral triangles.

Congruent shapes

2 These triangles are **congruent** because they remain the same size and shape even when they have been rotated.

Shade the congruent shapes.

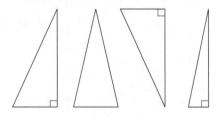

Similar shapes

3 These triangles are not congruent. They are the same shape but the sides are not the same length. They are similar because they have **congruent angles**.

Shade the three triangles that are similar because their angles are congruent.

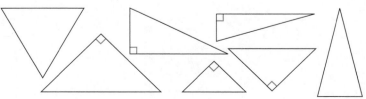

4 There are several types of quadrilaterals. Label these quadrilaterals. Use the word bank to help with spelling.

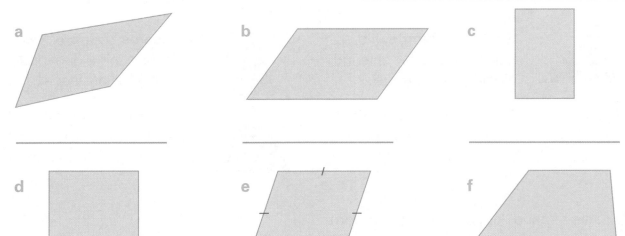

a _____

b _____

c _____

d _____

e _____

f _____

5 Write down something that is the **same** and something that is **different** about each pair of polygons.

Polygon pair	Something the same about the pair	Something different about the pair
e.g.	They both have 4 sides and 4 right angles. Both polygons have 2 pairs of parallel lines.	One has sides that are all the same length. The other has opposite sides that are the same length.
a		
b		
c		

Extended practice

1 Identify each polygon from its description.

    a This polygon has three sides, one right angle and two equal angles.

       It is _____.

    b This polygon has six equal angles. It is _____.

    c This polygon has four sides. It has one pair of sides that are parallel. It has another

       pair of sides that are not parallel. It is _____.

    d This polygon is a parallelogram. It has two acute angles and two obtuse angles.

       It has four equal sides. It is _____.

2 Write your own description of a polygon. Describe it accurately—but without making it too easy for someone to guess.

_____

_____

3 Write down some information about this polygon. _____

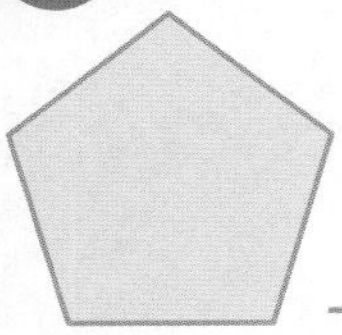

_____

_____

_____

4 This picture is made from:
- two right-angled triangles
- an irregular pentagon
- a trapezium
- a rectangle.

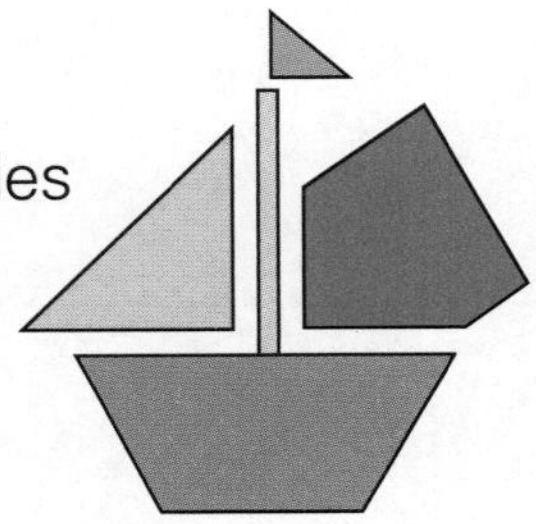

Draw a polygon picture. Write the names of the polygons that you use. (Remember: a polygon has no curved sides!)

A 3D shape has height, width and depth. A polyhedron is a 3D shape that has flat faces. A cube is a polyhedron but a cylinder is not. (The plural of polyhedron is *polyhedra*.)

Cube (a polyhedron)

Cylinder (**not** a polyhedron)

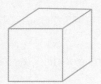

I am a 3D shape, but I am not a polyhedron!

Guided practice

1 Prisms and pyramids are two types of polyhedra. They get their names from the shapes of their bases. Use the word bank to help you write the names of these polyhedra.

Word bank

triangular pyramid   rectangular prism

hexagonal prism   octagonal prism

square pyramid   triangular prism

pentagonal prism

e.g. **hexagonal pyramid**

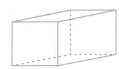

e.g. **square prism**

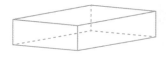

a _____

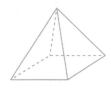

b _____

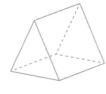

c _____

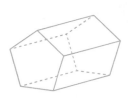

d _____

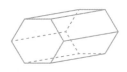

e _____

f _____

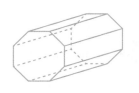

g _____

2 The side faces of prisms are always rectangles. What 2D shape can you see on the side faces of all pyramids? _____

OXFORD UNIVERSITY PRESS

1 Complete these sentences.

a I know this is a polyhedron because _____

_____.

b I know this is **not** a polyhedron because _____

_____.

2 Write the number of faces, edges and vertices on these 3D shapes.
You could use actual 3D shapes to help with this activity.

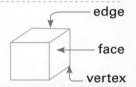

	3D shape	Number of faces	Number of edges	Number of vertices	Name of 3D shape
e.g.		7	12	7	Hexagonal pyramid
a					
b					
c					
d					
e					

A **pyramid** has one base. It usually sits on its base. A **prism** has two bases. A prism often sits on one of the side faces and not on the base.

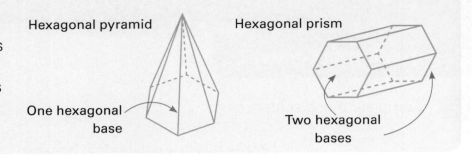

Hexagonal pyramid

One hexagonal base

Hexagonal prism

Two hexagonal bases

3 Complete this table.

	3D shape	Number of bases	Base shape	Side face shape	The object is sitting on:
e.g.	Hexagonal pyramid	1	hexagon	triangles	the base
a	Square pyramid				
b	Triangular prism				
c	Triangular pyramid				
d	Rectangular prism				

4 A rectangular prism would open out to make a net like this: For which 3D shapes are these the nets?

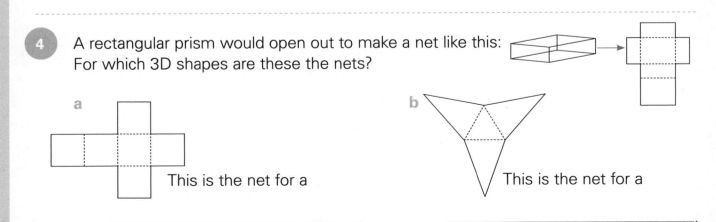

a

This is the net for a

_____.

b

This is the net for a

_____.

OXFORD UNIVERSITY PRESS

1 Drawing 3D shapes is difficult, because you have to make a 2D drawing look as though it has depth. Try to draw these objects on the isometric grid. The dotted lines show the "hidden" edges. It might take a few tries to make them look right.

a This is a:

_____

_____.

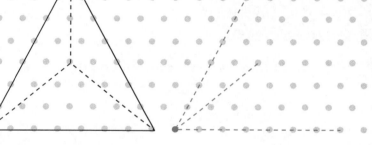

b This is a:

_____

_____.

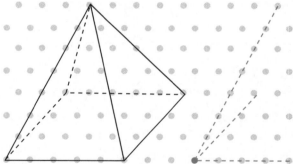

c This is a:

_____

_____.

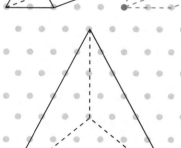

d This is a:

_____

_____.

2 If you made a cross-section of a cone in this direction, you would see a circle.

What 2D shape would you see if you cut across each object in the direction of the arrow?

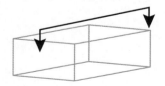

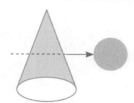

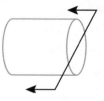

a _____   **b** _____   **c** _____

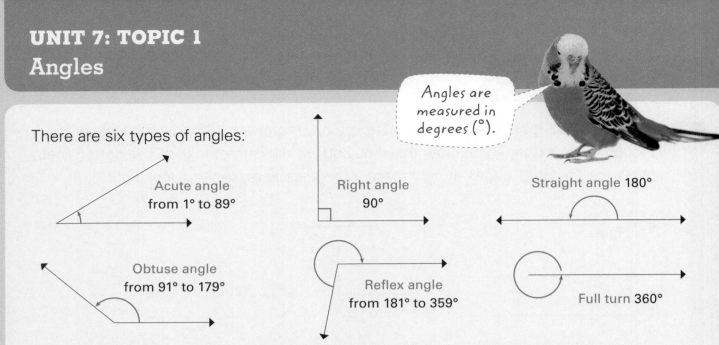

Angles are measured in degrees (°).

There are six types of angles:

Acute angle
from 1° to 89°

Right angle
90°

Straight angle 180°

Obtuse angle
from 91° to 179°

Reflex angle
from 181° to 359°

Full turn 360°

Perpendicular lines meet at a right angle. The lines on the right angle above are perpendicular to each other.

Guided practice

1 Write the name of each type of angle.

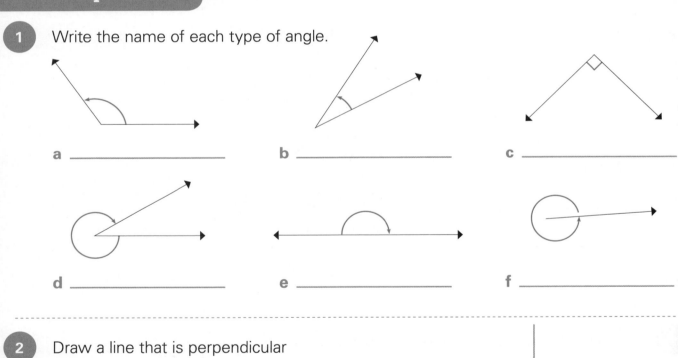

a _____

b _____

c _____

d _____

e _____

f _____

2 Draw a line that is perpendicular to each green line.

3 Use a pencil and ruler to draw each angle from the dot on its base line.

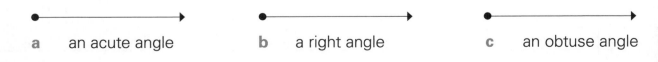

a   an acute angle

b   a right angle

c   an obtuse angle

Angles are measured with a protractor. The base line of the protractor needs to be on the base line of the angle. You have to make sure you read the correct track. This angle is on the inside track:

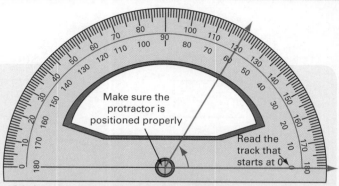

Make sure the protractor is positioned properly.

Read the track that starts at 0

1 Write the type and size of each angle.

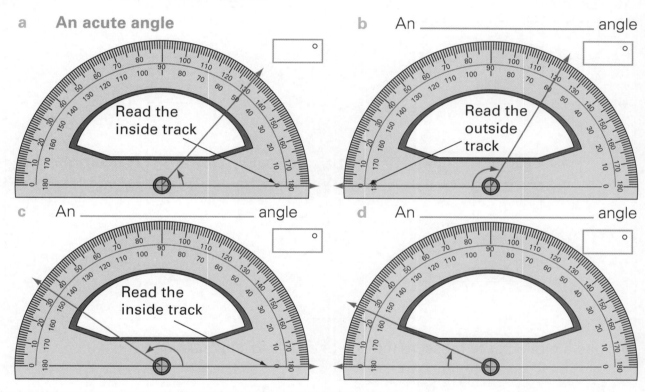

a An acute angle

Read the inside track

b An _____ angle

Read the outside track

c An _____ angle

Read the inside track

d An _____ angle

2 Write the type of angle. Circle the best estimate for the size of the angle.

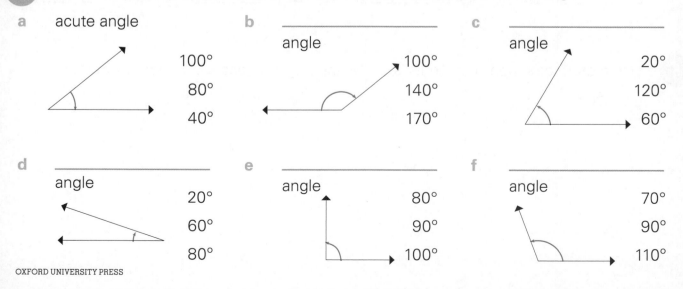

a acute angle

100°
80°
40°

b _____ angle

100°
140°
170°

c _____ angle

20°
120°
60°

d _____ angle

20°
60°
80°

e _____ angle

80°
90°
100°

f _____ angle

70°
90°
110°

3 Write an estimate for the size of each angle. Think about the type of angle. You could also think about how the size compares to a right angle.

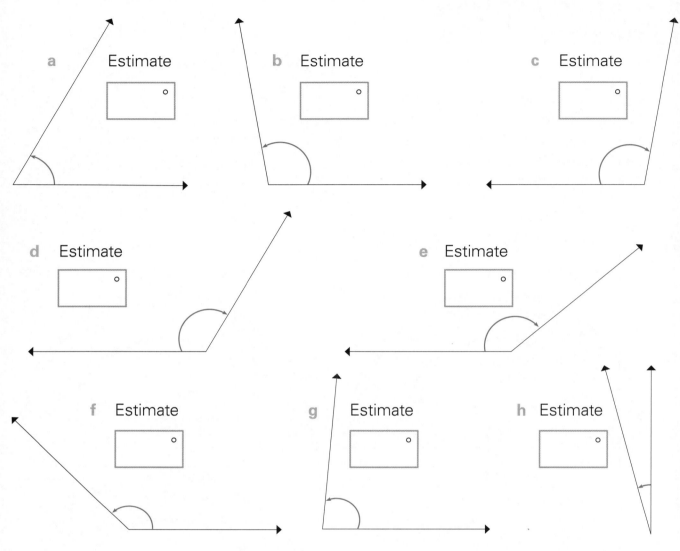

a Estimate [ ]°

b Estimate [ ]°

c Estimate [ ]°

d Estimate [ ]°

e Estimate [ ]°

f Estimate [ ]°

g Estimate [ ]°

h Estimate [ ]°

4 Use a protractor to measure the size of each angle in question 3.

a _____   **b** _____   **c** _____   **d** _____

e _____   **f** _____   **g** _____   **h** _____

5 Use a protractor, pencil and ruler to draw the angle on each line. Start at the dot.

a 70°   **b** 115°

OXFORD UNIVERSITY PRESS

Extended practice

This diagram shows one strategy you can use
to find the size of a reflex angle.

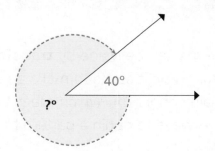

1 Without using a protractor, write the size
of the reflex angle in this diagram. _____

2 Use a strategy of your choice to find the size of these reflex angles.

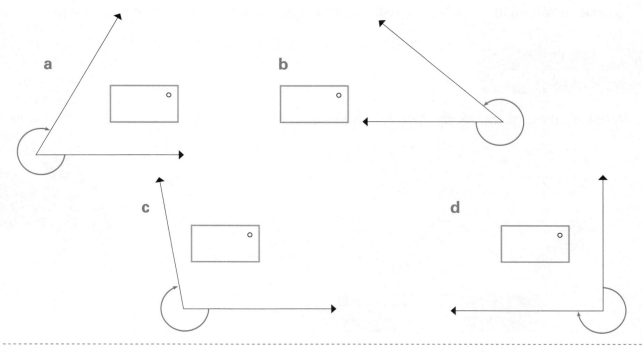

a

b

c

d

3 There are two angles
shown here.

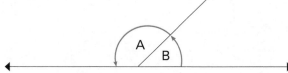

a Estimate the size of each angle.

Angle A estimate: ☐

Angle B estimate: ☐

b Explain how you estimated the size of each angle.

_____

c Measuring just **one** of the angles, write the sizes of **both** angles.

Angle A = _____   Angle B = _____

d Explain how you found the size of the angle that you **did not** measure.

_____

Patterns can be made by **transformation**. This means that as you move a shape in a certain way, it starts to make a pattern. When the pattern is formed, the shapes must remain congruent, which means that they are always the same shape and size. Here are some ways to begin a pattern by transforming a shape:

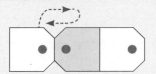
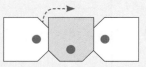
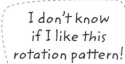

Translation (sliding it)   Reflection (flipping it over)   Rotation (turning it)

Guided practice

I don't know if I like this rotation pattern!

1 What method of transformation has been used?

a 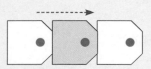 _____

b _____

c _____

---

2 Complete the patterns. Remember to keep the shapes congruent.

a Rotate the triangle.

b Translate the triangle.

c Reflect the triangle.

d Make a pattern of your choice.

e How did you transform the pentagon?

_____

Patterns can be made by transforming shapes **horizontally**, **vertically** or **diagonally**.

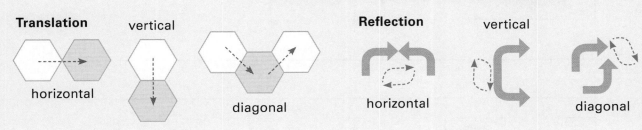

Translation vertical

horizontal

diagonal

Reflection vertical

horizontal

diagonal

1 Describe these patterns.

	Pattern	Description
e.g.		The triangle has been reflected vertically.
a		
b		
c		
d		
e		
f		

2 Continue this pattern and describe the way it grows.

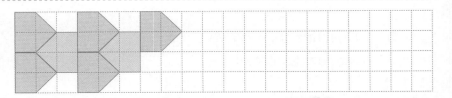

3 Look at the way these patterns grow. Complete each pattern, then describe it.

a

_____

_____

b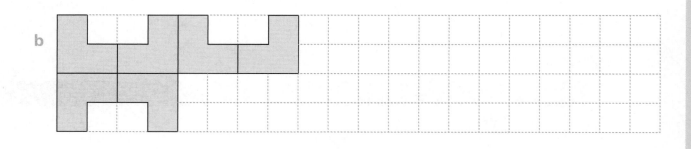

_____

_____

c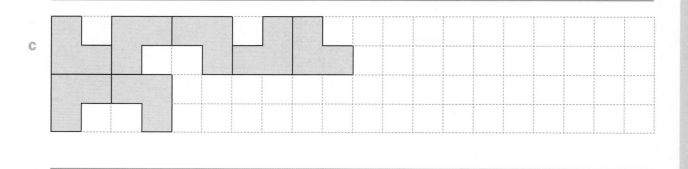

_____

_____

4 **a** Design a transformation pattern using this shape.

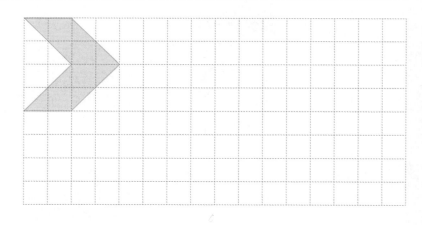

b Describe the way you made your pattern.

_____

_____

OXFORD UNIVERSITY PRESS

Extended practice

1 You can create designs in a few minutes with the help of a computer and a program such as Microsoft Word (or similar).

a Open a blank document. Make sure you can see the **Drawing** menu bar. If you cannot see it, click on **View**, then **Toolbars**, then **Drawing**.

b Click on the **Autoshapes** icon and choose an interesting shape.

c Draw the shape at the top of the page by clicking and dragging.

d Copy the shape.

e Paste the shape.

f Use the arrow keys to move the shape so that its left edge joins the right edge of the first shape, like this:

g Repeat steps d–f as many times as you like.

2 This activity involves rotating copies of a simple shape on top of the original shape.

a Open a new blank document.

b Click on the **Autoshapes** icon and choose a double arrow.

c Draw the arrow on the page by clicking and dragging.

d Copy and paste the shape as you did in question 1.

e Use the arrow keys to move the shape so that it is exactly over the top of the first shape.

f Select the shape to change it. (PC users: right-click and choose **format Autoshape**. Mac users: hold **control** as you click and choose **format Autoshape**.)

g Click the **size** tab and look for the **rotation** menu.

h Change the **rotation** amount from 0° to 30° and click **OK**.

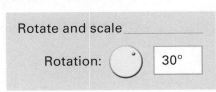

Rotate and scale

Rotation: 30°

i Copy, paste and move the new shape by repeating steps e–h.

j Repeat, increasing the angle of rotation by 30° each time.

There are two types of symmetry: **line** (mirror) symmetry and **rotational** (turning) symmetry. Some shapes have both line symmetry and rotational symmetry.

Line symmetry
One side is the same as the other.

Rotational symmetry
It fits on top of itself before it gets back to the starting point.

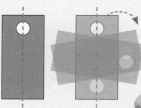

Line symmetry and rotational symmetry

Some shapes don't have any lines of symmetry.

Guided practice

1 Tick the shapes that have line symmetry.

A 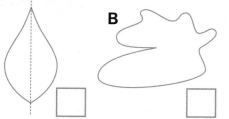 □
B □
C □
D □
E □

F 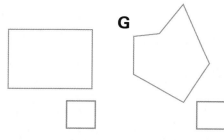 □
G □
H □
I □
J □

- -

2 All of the following shapes have line symmetry.

   a Draw at least one line of symmetry on each shape.

   b Some of the shapes also have rotational symmetry. Colour the shapes that have rotational symmetry.

A   B   C   D   E

F   G   H   I   J

Independent practice

Shapes can have more than one line of symmetry.
The red dotted lines show that this shape has two lines of symmetry.

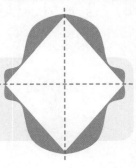

1 All these shapes have line symmetry. Use a strategy of your choice to find and draw in the lines of symmetry. Some have one, some have two—and some have four!

a 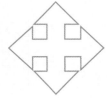   b 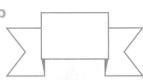   c    d

e    f 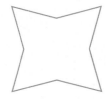   g    h

i    j 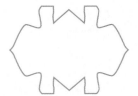   k 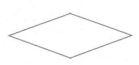   l

2 All regular 2D shapes have lines of symmetry. A triangle has three lines of symmetry. Identify and draw the lines of symmetry on these regular shapes.

a 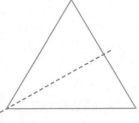   b    c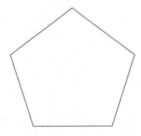

d 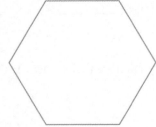   e 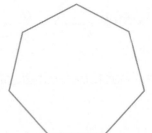   f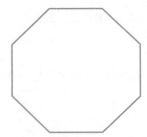

This shape has rotational symmetry of "order 2". That means that the shape fits on top of itself two times as it rotates, counting the starting position as one.

1st position

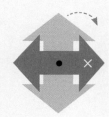

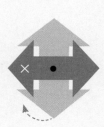

Back to the start

2nd position

3 Find the order of rotational symmetry for these shapes. You may wish to trace over the shapes and use cut-outs for this activity.

a

Rotational symmetry of order _____

b

Rotational symmetry

of order _____

c

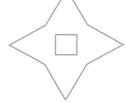

Rotational symmetry

of order _____

d

Rotational symmetry

of order _____

e

Rotational symmetry

of order _____

f

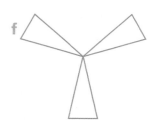

Rotational symmetry

of order _____

g

Rotational symmetry

of order _____

h

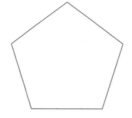

Rotational symmetry

of order _____

4 True or false? Every symmetrical shape has rotational symmetry of at least order 1.

_____

OXFORD UNIVERSITY PRESS

Extended practice

1 Some of the digits that make up our number system are symmetrical.
However, this depends on the way they are drawn.

1 2 3 4 5 6 7 8 9 0

a Draw the lines of symmetry to show any digits that are symmetrical.

b One of the 10 digits that can be drawn symmetrically is **not**
drawn symmetrically in the list. Which one is it? Re-draw it
and draw the line of symmetry.

c One of the 10 digits can be drawn so that it has an infinite
number of lines of symmetry. Re-draw it so that it has an
infinite number of lines of symmetry.

---

2 **a** You probably know some capital letters have lines of symmetry, such as
capital **A**. What is another capital letter that has one line of symmetry?

b The letter **S** has no lines of symmetry but it has rotational symmetry.
What is the order of symmetry for a capital **S**?

c What is another capital letter with rotational symmetry?

d Some capital letters have both line symmetry and rotational symmetry.
For example, a capital **H** has two lines of symmetry and has rotational symmetry
of order two. Complete the Venn diagram, showing the letters that have:

- line symmetry
- rotational symmetry
- both line symmetry and rotational symmetry.

Letters with line symmetry	Letters with both	Letters with rotational symmetry
A	H	S

---

3 We often see symmetry in nature. Or do we?
Look closely at this leaf. What, if anything,
makes it asymmetrical?

_____

When you enlarge something, you make it bigger. There are two simple ways of enlarging a 2D shape using a grid of squares. You can draw the picture on bigger squares, or you can increase the length of every line by the same amount.

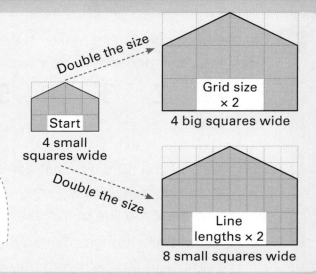

Double the size

Start
4 small squares wide

Grid size × 2

4 big squares wide

Double the size

Line lengths × 2

8 small squares wide

You can reduce the size of a picture by doing the opposite of the enlargement process.

Guided practice

1 Enlarge these shapes by re-drawing them on the larger grids.

a
b
c

d
e
f

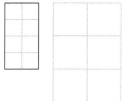

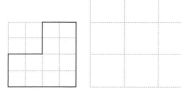

2 Enlarge these shapes by doubling the lengths of all the lines. Start each drawing at the red dot.

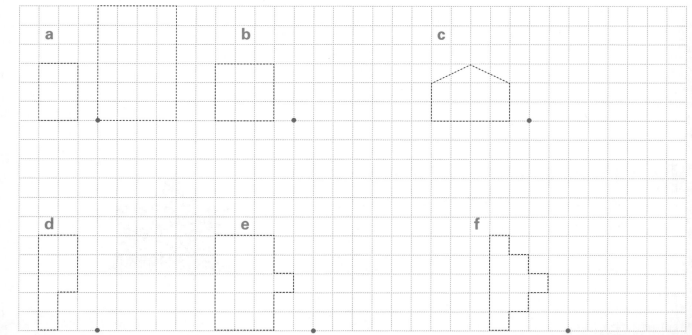

a
b
c

d
e
f

OXFORD UNIVERSITY PRESS

1 Enlarge each picture by drawing it on the larger grid.

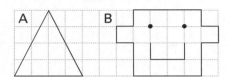

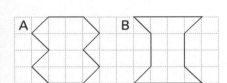

2 Draw an enlargement of each shape on the second grid. Then make an even bigger enlargement by drawing them on the third grid.

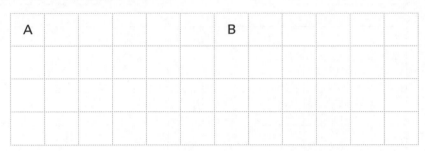

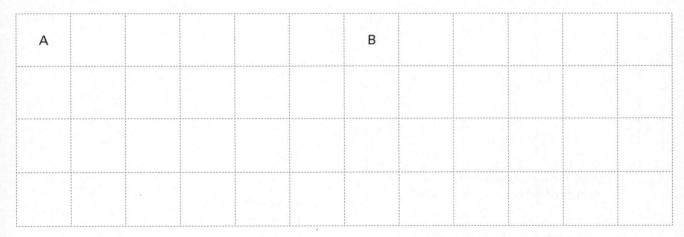

3 Reduce the size of these letters by drawing them on the smaller grid.

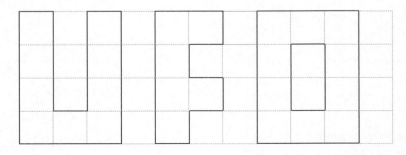

You can enlarge or reduce a picture by a **scale factor**. If you want a picture to be three times as big, you enlarge it by a scale factor of three.

4 Re-draw these pictures according to the scale factor shown. Start at the red dot.

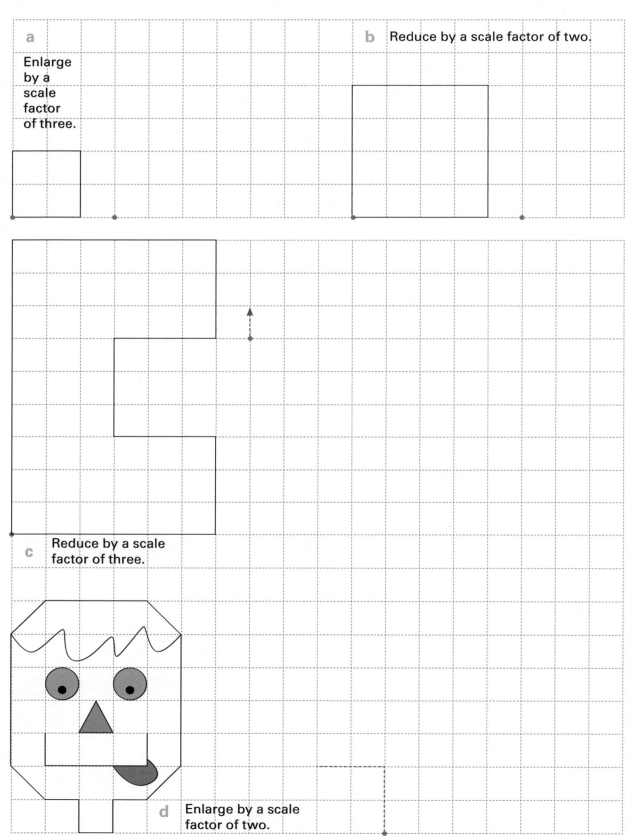

a Enlarge by a scale factor of three.

b Reduce by a scale factor of two.

c Reduce by a scale factor of three.

d Enlarge by a scale factor of two.

OXFORD UNIVERSITY PRESS

Extended practice

1 A 2 cm × 2 cm square has an area of 4 cm². If you enlarge a 2 cm square by a scale factor of two, what happens to the area? Experiment on a piece of spare paper before writing the answer.

_____

2 You will need access to a computer for the next activity. In a program such as Microsoft Word, you can enlarge a picture by clicking and dragging. You can enlarge pictures more accurately by changing the size of the picture by a percentage amount.

   a Open a blank Word document.

   b Insert a picture.

   c Select the picture to format it. (PC users: right-click and choose **format picture**. Mac users: hold **control** as you click and choose **format picture**.)

   d Click the **size** tab and look for the **scale** menu.

   e Change the **scale** amount from 100% to 200%. (If you click the **lock aspect ratio** button, this will change the width and height by the same amount.)

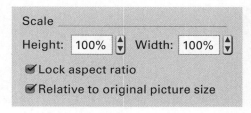

Scale _____
Height: 100% ⬍ Width: 100% ⬍
☑ Lock aspect ratio
☑ Relative to original picture size

   f Click **OK** and watch the picture change.

3 **a** If you were formatting the size of a picture in Word and you clicked "100%", what would happen to the size of the picture?

                                                            _____

   b How would you enlarge a picture by a scale factor of three in Word?   _____

   c Insert another picture and find a way to reduce the picture to half its size. How did you do it?   _____

4 If you do not lock the aspect ratio in Word, the height and width of a picture can be changed separately. This can make the pictures look strange but it can be fun to do. Try enlarging the width and height of a picture by different amounts.

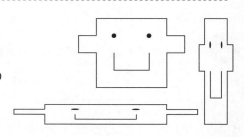

UNIT 8: TOPIC 4
Grid references

Grid references are a way of describing position. Grid references can mean the area inside a square or an exact point on the grid.
The circle is at B1 on both grids.
On a grid like Grid B, there can only be one object at each point.

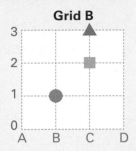

Grid A

Showing a position inside a square.

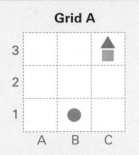

Grid B

Showing a position at an exact point.

To read a coordinate point, first go ACROSS the river then UP the mountain!

Guided practice

1 What is the position of these shapes on the grids at the top of the page?

Grid A square: _____   Grid B square: _____

Grid A triangle: _____   Grid B triangle: _____

---

2 What are the positions of these shapes on Grid C?

a The diamond:

_____

b The star:

_____

c The triangle:

_____

d The circle:

_____

Grid C

---

3 Draw the following on Grid C: **a** The letter O at A4   **b** A smiley face at B3

c The letter K at B4   **d** An oval at D2

e The letter R at B5   **f** The letter U at A5

---

4 Which shape has a different grid reference on Grid C and Grid D? _____

Grid D

---

5 Draw the letter ✖ at the following points on Grid D:
B4   C3   D2   E1

---

6 Write a coordinate point that would continue the ✖s in a diagonal line. _____

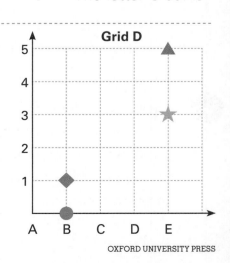

Independent practice

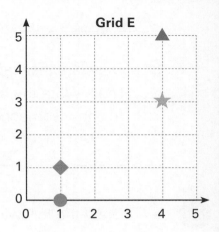

Grid E

Coordinate points are often written with two numbers, rather than a letter and a number. You read the number going across first, then the number going up. The numbers are in brackets, separated by a comma. The circle is at (1,0).

1 Write the grid references for:

a the star _____ b the triangle _____ c the diamond _____

2 Draw the following on Grid E. a a square at (2,3)

b circles at (1,4) and (3,4) c stars at (1,2), (2,2) and (3,2)

3 a Write the first letter of your first name on an empty grid point.

b What is the grid reference for the letter you wrote? _____

4 When two coordinate points have an arrow between them, it means that you join them with a straight line. Complete the following.

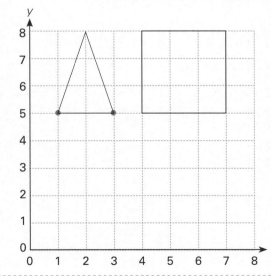

a The coordinate points for drawing the triangle

are: (1,5) ⟷ (3,5) ⟷ (2,8) ⟷ _____.

b The coordinate points for drawing the square

are: (4,5) ⟷ _____ ⟷ _____ ⟷

_____ ⟷ _____.

5 a Draw a large rectangle on the grid lines below the triangle and the square.

b Write the coordinate points for drawing the rectangle.
Remember to end the drawing back at the starting point. _____

_____

6 **a** Write the coordinate points to show someone how to draw this letter N.

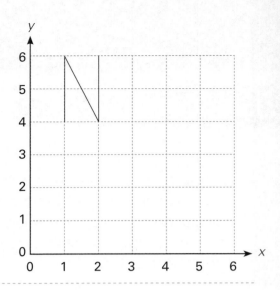

_____

b Draw another capital letter using straight lines. Write the coordinate points to show someone how to draw the letter.

_____

7 **a** Draw the following picture by plotting with dots and joining the coordinate points.

(1,1) ⟷ (4,1) ⟷ (4,4) ⟷ (5,4) ⟷

(5,1) ⟷ (8,1) ⟷ (6,2) ⟷ (6,6) ⟷

(8,4) ⟷ (8,5) ⟷ (6,7) ⟷ (5,7) ⟷

(5,8) ⟷ (7,8) ⟷ (7,9) ⟷

(8,9) ⟷ (8,10) ⟷ (7,10) ⟷ (7,11)

⟷ (6,12) ⟷ (3,12) ⟷ (2,11) ⟷

(2,10) ⟷ (1,10) ⟷ (1,9) ⟷ (2,9)

⟷ (2,8) ⟷ (4,8) ⟷ (4,7) ⟷ (3,7)

⟷ (1,5) ⟷ (1,4) ⟷ (3,6) ⟷ (3,2)

⟷ (1,1) STOP!

b Draw a mouth:
(3,10) ⟷ (3,9) ⟷ (6,9) ⟷ (6,10).

c Draw a nose: (4,10) ⟷ (5,10), then draw an oval shape around the line.

d Draw two eyes. What are the coordinate points?

_____

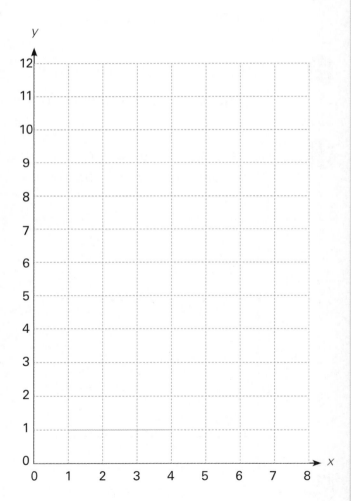

OXFORD UNIVERSITY PRESS

Extended practice

1 Design a coordinate picture on the grid. Afterwards, you can give the instructions to someone so that they can draw the picture. Some things to think about:

- Don't make the picture too complicated.
- Make sure the coordinate points are correct. (If you can't follow them, nobody else will be able to!)
- Use straight lines where possible.

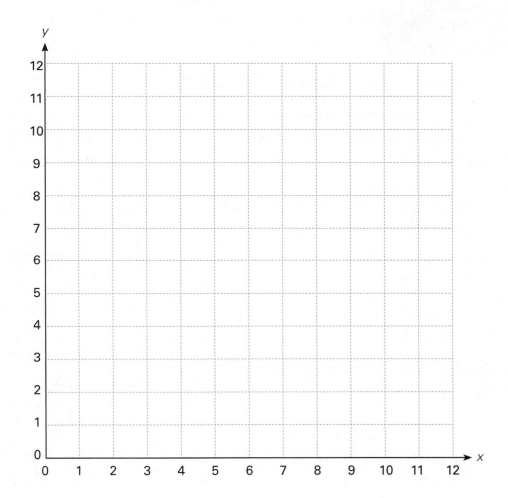

2 Write the instructions in the way that they were written on page 122.
Try following your instructions on a grid before you give them to somebody else.

_____

_____

_____

The four main compass directions are north, south, east and west. To remember their position on a compass rose, some people use sayings such as "never eat slimy worms".

We need a compass direction to describe the position of the dog.

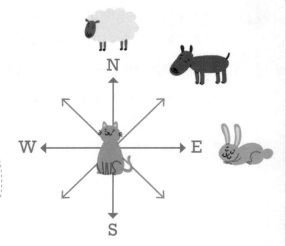

Guided practice

1. The dog on the compass rose is north-east of the cat.

   a. Label the four empty arrows NE (north-east), SE (south-east), SW (south-west) and NW (north-west).

   b. Draw a triangle at the SW point, a circle at the NW point and a square at the SE point.

2. The teacher is at the centre of this classroom. Use the plan to find the answers.

   a. Who is north-east of the teacher? _____

   b. Who is south-west of the teacher? _____

   c. Use a compass direction to describe Sam's position.

   _____

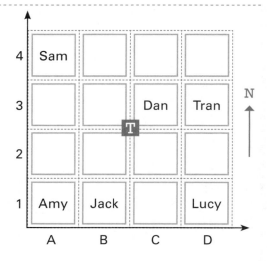

3. It is not possible to describe the position of Jack's table using one of the eight compass directions, but we can use a grid reference. Jack is at B1.

   a. Use a grid reference to describe the position of Tran. _____

   b. Eva is at A2. Write her name on the plan.

   c. Choose a position to the east of Sam and write your initials in it.

   d. What is the grid reference for the position? _____

   e. Write Jo's name in a position between Sam and Lucy. Use a compass direction to describe the position in relation to the teacher and a grid reference for the table. _____

OXFORD UNIVERSITY PRESS

Independent practice

Use this map of Jo's town for the following activities.

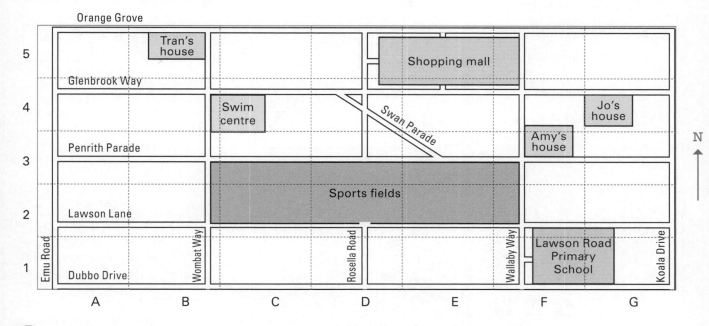

1

a In which direction would Jo go to get from home to the swim centre? _____

b What is the grid reference for Tran's house? _____

c Imagine you lived at the **north** end of Rosella Road. Shade in the shortest route along the roads that would take you to the southern entrance of the sports fields.

d Amy lives on Penrith Parade. Using compass directions and street names, write instructions to get from Amy's house to the swim centre.

_____

_____

e True or false? Amy's house is north-east of Jo's house. _____

f The Magic Movie Theatre is **north** of the Swim centre, on the **right-hand side** of Wombat Way. Draw and label it on the map.

g Write the grid reference for the north-eastern corner of Lawson Road Primary School. _____

h Using compass directions and street names, write instructions to get from Tran's house on Wombat Way to the school entrance on Wallaby Way.

_____

_____

2 A legend (or key) gives information about places on a map. If a map has a scale, you can work out real-life distances. On this map, the treasure is at (4,1) and is 50 metres from the snake pit.

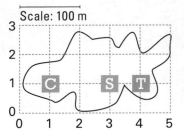

Legend
T = Treasure
S = Snake Pit
C = Campsite

a Write a coordinate for the campsite.

b How far is the campsite from the treasure?

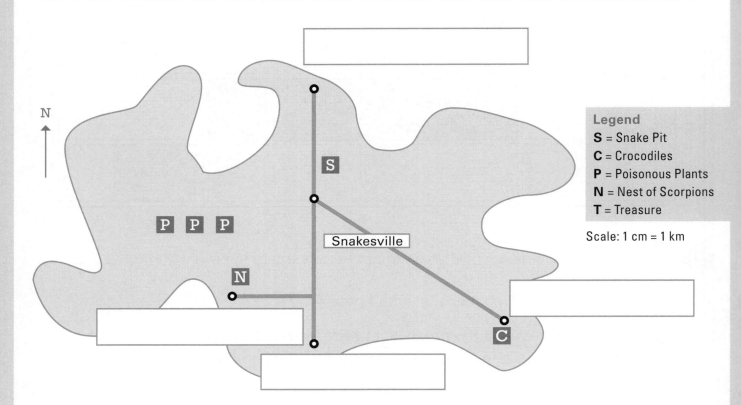

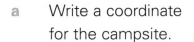

N

Legend
S = Snake Pit
C = Crocodiles
P = Poisonous Plants
N = Nest of Scorpions
T = Treasure

Scale: 1 cm = 1 km

3 **a** Use the legend to mark these on the map: Big Bug Beach has a nest of scorpions near it. Tin Pot Cave is 6 km south-east of Snakesville. Spider Head is 4 km south of Snakesville. Cockroach Cliff is 7 km north of Spider Head.

b Shark Point is 5 km west of Snakesville. Mark its position with a dot and write "Shark Point" on the map.

4 **a** There is a curved track from Snakesville to Shark Point that misses the poisonous plants by going to the **south** of them. Draw the track on the map.

b Estimate the distance from Snakesville to Shark Point along the track you drew.

5 **a** Goanna Gorge is 2 km to the north-east of Spider Head. Mark it on the map with a dot and the letter G. Draw a straight road from Spider Head to Goanna Gorge.

b The Treasure is buried along a straight track 500 m south-east of Goanna Gorge. Mark it on the map with the letter T.

OXFORD UNIVERSITY PRESS

Extended practice

1 CAT stands for "Computer Artist's Toy". It moves according to directions given and traces a path with a pen. The CAT needs to be programmed to draw an octagon. The first two moves are shown. CAT is programmed to understand distance and compass directions.

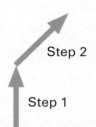

Step 2

Step 1

a Write the steps that would complete the octagon.

Step 1. Move north 2 cm.

Step 2. Move north-east 2 cm. _____

_____

_____

b Follow your own directions to see if you draw an octagon. If you want your octagon to be accurate, you will need to use a protractor. Or your teacher may ask you to use a line tool in a computer program such as Microsoft Word.

2 Draw your own Treasure Island Map. Include a legend for some interesting places on the island, a scale and a direction indicator. Label the grid so that you can give grid references for the places on the map.

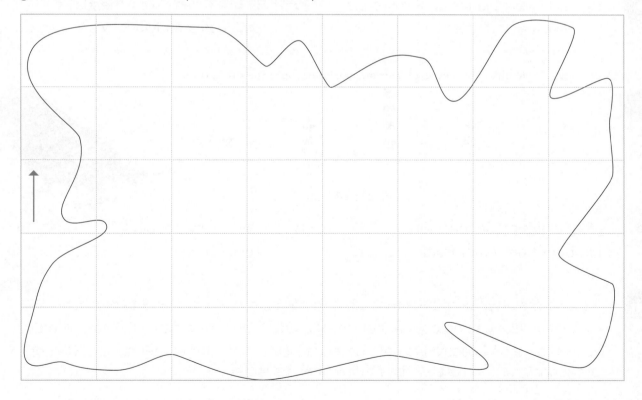

Legend		Scale:

A common way to represent data is on a graph. There are several types of graph. The type of graph used depends on what is being represented.

Graph to show our favourite snack

- fruit
- chocolate
- ice-cream

Guided practice

Vertical or horizontal bar graph

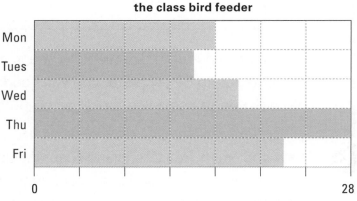

Number of birds that visited the class bird feeder

Number of birds that visited the class bird feeder

1

a Fill in the blanks on the number axis on each graph.

b By how many was Tuesday's total less than Monday's? _____

Dot plot

2

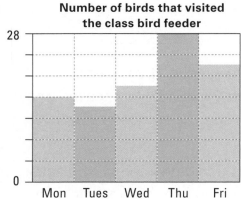

Number of pieces of fruit our group brought for snack time

Pieces of fruit

a What is the most common number of pieces of fruit? _____

b How many people were surveyed? _____

The graphs in questions 1 and 2 show numerical data.

3 The two main types of data that are collected are *numerical* and *categorical*. Numerical data can be counted (or measured). Categorical data (such as where we like to go on holidays) is not numerical. Write "N" (for numerical) or "C" (for categorical) for the type of data that will be collected.

a What is your favourite pet? _____ **b** How many pets do you have? _____

c How tall are you? _____ **d** What is your favourite sport? _____

e What is your favourite subject? _____ **f** How long do you spend reading each day? _____

Independent practice

1 If you asked, "How many snacks do you eat a day?", you would be collecting numerical data. Write a survey question about food that would enable you to collect categorical data.

_____

2 If you asked, "What type of music do you like?", you would be collecting categorical data. Write a survey question about music that would enable you to collect numerical data.

_____

3 Class 5T took the noon temperature for 20 days:

19°, 18°, 19°, 20°, 19°, 20°, 20°, 20°, 19°, 18°, 20°, 19°, 20°, 19°, 18°, 20°, 18°, 17°, 19°, 20°

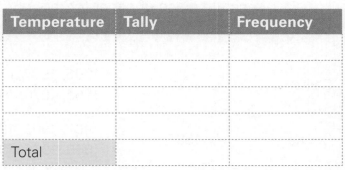

Temperature	Tally	Frequency
Total		

a What type of data did they collect?

_____

b Complete the frequency table for the data.

c Complete the dot plot for the data about the temperatures.

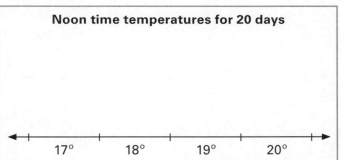

Noon time temperatures for 20 days

17°   18°   19°   20°

4 **a** Create a frequency table about the colour of people's hair in your class.

Frequency table showing hair colour

Colour	Light	Medium	Dark	Total
Frequency				

b Transfer the data onto a bar graph. Decide on a suitable scale.

c What other type of graph would also be suitable for this data?

_____

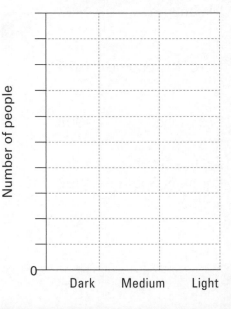

Hair colour in our class

Number of people

0

Dark   Medium   Light

5 Add to the information in question 4 by creating a two-way table showing the lengths and colours of students' hair in your class.

Hair type	Light	Medium	Dark	Total
Short length				
Medium length				
Long length				
Total				

6 This table shows the top ten premiership winning teams in the Australian Football League.

a Complete the total column.

b Decide on a suitable type of graph and scale to display the information. Use a separate piece of paper for this.

Club	Year started	Premiership years	Total
Carlton	1897	1906, 1907, 1908, 1914, 1915, 1938, 1945, 1947, 1968, 1970, 1972, 1979, 1981, 1982, 1987, 1995	
Collingwood	1897	1902, 1903, 1910, 1917, 1919, 1927, 1928, 1929, 1930, 1935, 1936, 1953, 1958, 1990, 2010	
Essendon	1897	1897, 1901, 1911, 1912, 1923, 1924, 1942, 1946, 1949, 1950, 1962, 1965, 1984, 1985, 1993, 2000	
Fitzroy	(1897–1996)	1898, 1899, 1904, 1905, 1913, 1916, 1922, 1944	
Geelong	1897	1925, 1931, 1937, 1951, 1952, 1963, 2007, 2009, 2011	
Hawthorn	1925	1961, 1971, 1976, 1978, 1983, 1986, 1988, 1989, 1991, 2008, 2013, 2014, 2015	
Melbourne	1897	1900, 1926, 1939, 1940, 1941, 1948, 1955, 1956, 1957, 1959, 1960, 1964	
North Melbourne	1925	1975, 1977, 1996, 1999	
Richmond	1908	1920, 1921, 1932, 1934, 1943, 1967, 1969, 1973, 1974, 1980, 2017	
Sydney Swans (formerly South Melbourne)	1897	1909, 1918, 1933, 2005, 2012	

OXFORD UNIVERSITY PRESS

Extended practice

Researchers believe that 10-year-old children have a vocabulary of 10 000 words, but it is very difficult to collect reliable data about the number of words anyone knows.

1   Without doing any research, write down three words that you think are used a lot

in everyday writing. _____

2   Do some research to see if you are right. You need 100 words of a text.

   a   Skim through and make a mental note of any words that you think are used frequently.

      •   Write these common words down.

      •   Do an accurate tally of the number of times the words are used.

   b   Which three words are most commonly used? _____

   c   Compare your research with that of somebody else. How does it compare?

   _____

   _____

3   There are a lot of vowels used in the 40 words of this joke.

   A monkey goes into a café and points to a picture of a cheese sandwich.
   "That's strange!" says one waitress to another. "A monkey is ordering a cheese sandwich."
   "I know!" says the monkey. "I usually order a hot dog."

   a   Find out how often each vowel is used. Make an accurate tally of the number for each vowel.

Vowel	A	E	I	O	U
Frequency					

   b   How reliable do you think this data is as an indicator of the most frequently used vowels? Why?

   _____

   _____

   _____

Representing and interpreting data

Two types of graphs used to represent data are **line graphs** and **circle graphs**.

A line graph is used to show how something changes over time, such as the amount of money in a piggy bank.

How much money was in Tran's piggy bank?

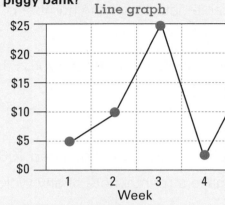

Line graph

Favourite colours of our class

Circle graph

A circle graph is a quick way to show small amounts of data.

Guided practice

1 The line graph above shows that the amount of money Tran had in week 1 was $5.

a By how much did it go up in week 2? _____

b In which week did Tran have the most money? _____

c Estimate the amount of money Tran had in week 4. _____

2 **a** Circle the correct statement about the circle graph above.

- Yellow is more popular than red.
- Blue is the least popular colour.
- Out of the 24 students in the class, 10 students chose blue.

You can only make an estimate of the numbers shown in a circle graph.

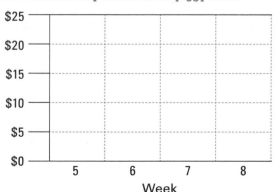

b Estimate the number of students who chose green. _____

3 These are the amounts Tran had in weeks 5–8. Use the information to make a line graph.

- Week 5: $15
- Week 6: $20
- Week 7: $3
- Week 8: $16

How much money was in Tran's piggy bank?

OXFORD UNIVERSITY PRESS

Independent practice

1 These are Eva's spelling scores out of 20 during the term.

Represent the data on a line graph.

Week	1	2	3	4	5	6	7	8	9	10
Score	20	18	19	14	6	16	20	20	17	15

a Decide on a suitable scale for the vertical axis.

b Write a title for the graph.

c Write appropriate labels for the horizontal and vertical axes.

d Plot the data, then join up each point.

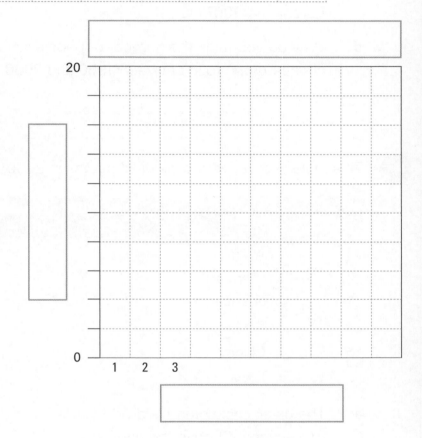

2 **a** In which weeks did Eva score 100%? _____

b Describe the change in scores between weeks 5 and 7.

_____

c In which week do you think Eva did not do her homework? _____

d True or false? Eva's average score was more than 16 out of 20. _____

e Between which weeks was the rise in scores the biggest? _____

3 The circle graphs show the top five holiday destinations for Australians in 1950 and 2000. The data was collected from a survey of 1000 people.

Favourite holiday destinations – 1950
- ■ New South Wales
- ■ Queensland
- ■ Victoria
- ■ New Zealand
- ■ Europe

a   What was the most popular destination in 1950? _____

b   The popularity of which place was the same in 1950 and 2000? _____

c   About how many people preferred to travel to Europe in 2000? _____

Favourite holiday destinations – 2000
- ■ Australia
- ■ Europe
- ■ Asia
- ■ New Zealand
- ■ USA

d   Why do you think the number of people who chose Europe rose between 1950 and 2000?

_____

4 This table shows the top six girls' names in 2000.

Rank	Name	Number of sections in circle graph	Key (colour used in circle graph)
1	Emily		
2	Ellie		
3	Jessica		
4	Sophie		
5	Chloe		
6	Lucy		

a   The blank circle graph is divided into 24 sections. Choose the number of sections to shade for each of the names in the table.

b   Choose a colour for each name and shade the graph. Then shade the key in the table.

c   Write a title for the circle graph.

d   Write a question that a teacher might ask Year 5 students about the information in the graph.

_____

_____

Title: _____

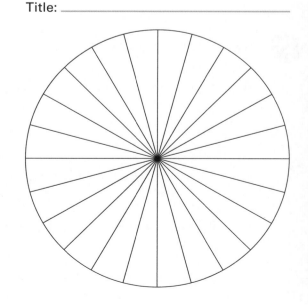

OXFORD UNIVERSITY PRESS

Extended practice

1 This information shows the number of points scored by each player (including the substitute) on a basketball team.

Player	Points in Game 1	Points in Game 2	Points in Game 3	Points in Game 4	Points in Game 5	Total number of points	Average points per game
Sam	17	19	19	14	16		
Amy	8	7	0	2	8		
Tran	5	8	4	2	11		
Eva	14	15	3	11	17		
Lily	2	4	1	0	3		
Noah	6	2	4	2	21		

a Divide the total for each player by the number of games to find their average number of points per game. Write the average scores in the table.

b Use the data to create a graph of your choice. You may use the scaffold if you wish. Examples of questions you could focus on include:

- What were the highest totals for each player?

- How did Eva's (or anyone else's) scores change over the five games?

- What did the average scores look like after three games?

c Whose average score was the highest?

_____

Title

d Who scored the most points in a single game? _____

e Which player do you think spent most time on the sideline? _____

Give a reason for your answer.

_____

If you guess heads or tails, there is just as much chance that you will be right as there is that you will be wrong.

Will it be heads or tails?

In words:
There is an even chance.

As a percentage:
There is a 50% chance.

As a fraction:
There is a $\frac{1}{2}$ a chance.

As a decimal:
There is a 0.5 chance.

Guided practice

Using the probability words *certain, likely, even chance, unlikely* and *impossible*, describe the chance of the following things happening.

1

a The voice I hear when I turn on the radio will be a woman's. _____

b A cow will read the news on TV tonight. _____

c Someone will fall over at lunchtime. _____

d Tuesday will follow Monday next week. _____

2 The chance words in question 1 can be put on a number line. Write the other four words on the line: *certain, likely, unlikely, impossible*. Draw arrows to the positions you think are appropriate.

Even chance

0   0.1   0.2   0.3   0.4   0.5   0.6   0.7   0.8   0.9   1

3 There is $\frac{1}{4}$ of a chance that the spinner will land on red.

Which fraction describes the chance of the spinner landing on blue?

4 This spinner has a 90% chance of landing on red.
What is the percentage chance of it landing on green? _____

5 The chance of this spinner landing on yellow is 0.1.
What chance is there that it will land on:

a blue? _____   **b** green? _____   **c** white? _____

OXFORD UNIVERSITY PRESS

Independent practice

1 Read the descriptions of the chance of these events occurring.

  a Convert the chance words to a decimal to describe the chance of the event occurring.

  b Fill in the gaps where necessary.

		Value
A:	It is *impossible* to run 100 metres in two seconds.	
B:	It is *almost impossible* for me to win ten million dollars.	**0.1**
C:	It is *likely* that I will see a movie at the weekend.	
D:	There is *a better than even chance* that I will like the movie.	
E:	It is *very likely* that	
F:	There is an *even chance* that the next baby born will be a girl.	
G:	There is *less than an even chance* that I will go swimming tomorrow.	
H:	It is *almost certain* that	
I:	It is *certain* that	
J:	It is *very unlikely* that	
K	It is *unlikely* that	

2 Place the letter for each description in question 1 at an appropriate place on this number line.

3 Which do you think is more accurate when describing chance situations: number values or chance words?

_____

4 Which spinners have the following chance of landing on blue?

  a 75% chance _____

  b 1 out of 2 chance _____

  c $\frac{1}{3}$ of a chance _____

  d 100% chance _____

A   B   C   D

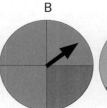 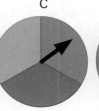 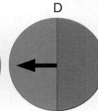

5 Colour this spinner so that the following probabilities are true:

- There is 0.1 chance for yellow.
- There is 0 chance for white.
- There is $\frac{2}{10}$ chance for blue.
- There is 0.4 of a chance for green.
- There is $\frac{3}{10}$ chance for red.

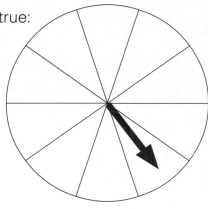

6 Each of these spinners can land on red, but there is not the same chance for each of them.

A   B   C   D   E

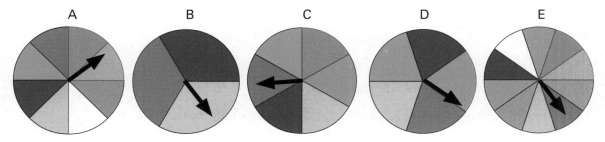

a   Order the spinners from **least likely** to **most likely** to land on red.

_____

b   Write a number value for the chance of each spinner landing on red.

Spinner A: Spinner B: Spinner C: Spinner D: Spinner E:

_____   _____   _____   _____   _____

7 Imagine this situation. There are 100 marbles in a bag. You know they are either red or blue. You pick out 10 marbles and find that you have 2 red ones and 8 blue ones.

How many of the 100 marbles are likely to be:

a   red?   _____   b   blue?   _____

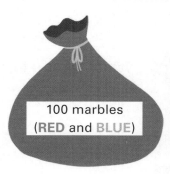

100 marbles
(**RED** and **BLUE**)

8 Which of these does **not** show the chance of choosing a blue marble from this bag? Circle one.

$\frac{1}{4}$ $\frac{4}{10}$ 40%   0.4

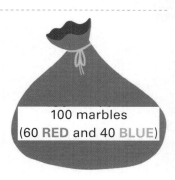

100 marbles
(60 **RED** and 40 **BLUE**)

OXFORD UNIVERSITY PRESS

Extended practice

1 In a pack of 52 playing cards (without the joker), there are four suits (or types): diamonds, spades, clubs and hearts. Imagine the 52 cards are face-down.

    a Express the chance of picking up a diamond as a fraction. _____

    b Name a type of card that you would have half a chance of picking up. _____

    c There are four "picture" cards in each suit. Express the chance of picking up a picture card as a fraction. _____

    d If you picked up 20 cards, how many could you expect to be hearts? _____

----

2 Chloe invented a board game. The number of squares you move depends on the colour of the spinner you land on. The more squares you move, the less chance there is of landing on that colour. This is how it works:

- Land on red: Move 1 square
- Land on blue: Move 2 squares
- Land on green: Move 4 squares
- Land on gold: Move 6 squares

    a Colour the spinner so that there is the greatest chance of landing on red, less chance for green, even less chance for blue and the least chance of all for gold.

    b Describe the chance of landing on each colour as a fraction and a decimal.

Red: ☐ _____   Blue: ☐ _____   Green: ☐ _____   Gold: ☐ _____

----

3 The Jellybean Company always put 20 red ones, 10 green ones, 25 white ones, 20 yellow ones, 10 purple ones, 10 pink ones and 5 black ones in each pack.

    a Joel loves the yellow ones. He takes one from his pack without looking. What is the chance that he will take a yellow jellybean? _____

    b Which colour is there a quarter of a chance Evie will take out? _____

    c Lachlan's favourites are red and green. What fraction of a chance does he have of getting one of his favourites? _____

    d Which colour jellybean has a 1-in-20 chance of being chosen by Charlie? _____

There is a 1-in-2 chance of choosing correctly when a coin is tossed. That means that if you toss the coin four times the chances are that it will land twice on heads and twice on tails. However, does that mean it **will** happen?

1st toss   **2nd toss**   **3rd toss**   **4th toss**

It's sure to land on tails next time, isn't it?

Guided practice

1 Imagine a coin lands on heads 10 times in a row. Circle the chance of it landing on tails next throw.

100%   90%   75%   50%   25%   0%

2
a Predict the result if you toss a coin 10 times.   Heads: _____   Tails: _____

b Toss a coin 10 times. Record the results.

Toss	1st	2nd	3rd	4th	5th	6th	7th	8th	9th	10th
H or T?										

c Compare your prediction with what actually happened. Explain the difference.

_____

3 There is not a 1-in-2 chance of rolling a 4 on a 6-sided dice.

a Give a number value for the chance of the dice landing on 4: _____

b If a dice lands on 4 ten times in a row, what is the chance of it landing on 4 on the eleventh throw? _____

4
a Predict the result if you roll a dice 12 times.

One: _____   Two: _____   Three: _____   Four: _____   Five: _____   Six: _____

b Roll a dice 12 times. Record the results.

Toss	1st	2nd	3rd	4th	5th	6th	7th	8th	9th	10th	11th	12th
Result												

c Was it more difficult to predict the results for the coin or the dice? Try to give an explanation. _____

Independent practice

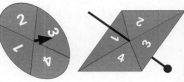

For this experiment, you will need a spinner numbered from 1 to 4.

1   **a**   Circle the number value that does **not** describe the chance of the spinner landing on number 4.

$\frac{1}{4}$ 4 out of 10   1 out of 4   25%   0.25

   b   If you spin the spinner four times it **should** land on each number once. Do you think that will happen? Give a reason for your answer.

_____

2   It's time to conduct the experiment. Decide on the number of spins that is necessary to obtain accurate results: 12? 20? 40? (The number needs to be a multiple of 4.) Operate the spinner and tally the results in the table.

Number on the spinner	1	2	3	4
Tally of the number of times it landed				
Total				

3   Write a few sentences about the results of the experiment. Think about things such as:

- Did it turn out how I expected?
- Why did it not land the same number of times on each number?
- If I started from the beginning again, would the results be the same?
- If I doubled the number of spins, would it be very different?
- How do my results compare to someone else's?

_____

_____

4   The way an experiment is set up can affect the results. If you made a 5-sided spinner and numbered it like this, how would it affect the chance for each number?

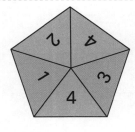
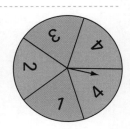

5 For this experiment, you will need two coins. There are three results that can occur. Fill in the table to show the possible results.

When you toss two coins the result can be:		
They both land on heads.		

6 Predict the results after 40 tosses of the coins.

Two heads: _____   Two tails: _____   Heads and tails: _____

7 Carry out the experiment. Tally and record the results in the table.

Ways the coins landed	Two heads	Two tails	Heads and tails
Tally of the number of times they landed like that			
Total			

8 Write a few sentences commenting on the results of your experiment.

_____

_____

_____

_____

9 Each result did not have the same chance of occurring in the last experiment. Explain why by looking at the diagram.

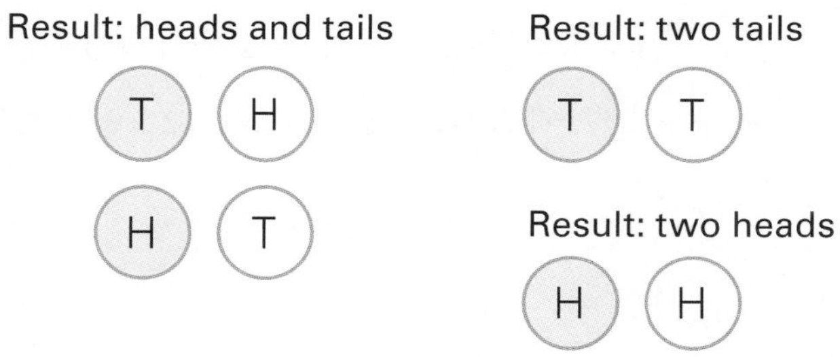

Result: heads and tails

T H

H T

Result: two tails

T T

Result: two heads

H H

_____

_____

OXFORD UNIVERSITY PRESS

Extended practice

1 We can give a number value for the chance of this spinner **not** stopping on white. Write the number value in as many different ways as you can.

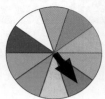

_____

2 Circle the statement that best describes the chance of this spinner stopping on yellow.

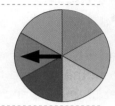

about 5%   about 15%   about 25%   about 50%   about 75%

3 Imagine the spinner in the question 2 stops on yellow 10 times in a row. Circle the chance of this spinner stopping on yellow next time.

$\frac{1}{6}$ $\frac{3}{6}$ 0   1

4 Draw seven equal squares on a piece of paper or card. Write the word MINIMUM, with one letter per square.

Cut out the seven squares. Turn them over. Move them around.

a What is the chance of picking up the letter **M** first go?   _____

b Have the letters **N** and **U** facing up and the other papers facing down. Describe the chance of picking up a letter **I** first go.   _____

c Put all the papers face-down and shuffle them around. Pick up two without looking. There might be two letter **M**s. What are all the possibilities?

_____

_____

_____

5 If you carried out the experiment 42 times, how many times would you expect each letter to appear?

_____

_____

_____

GLOSSARY

acute angle   An angle that is smaller than a right angle or 90 degrees.

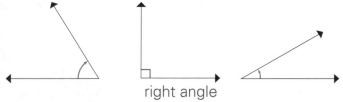

right angle

addition   The joining or adding of two numbers together to find the total. Also known as *adding, plus* and *sum*. See also *vertical addition*.

★★★ + ★★ = ★★★★★

3 and **2** is   **5**

algorithm   A process or formula used to solve a problem in mathematics.

Examples:
horizontal   vertical
algorithms   algorithms
24 + 13 = 37

	T	O
	2	4
+	1	3
	3	7

analogue time   Time shown on a clock or watch face with numbers and hands to indicate the hours and minutes.

angle   The space between two lines or surfaces at the point where they meet, usually measured in degrees.

75-degree angle

anticlockwise   Moving in the opposite direction to the hands of a clock.

area   The size of an object's surface.

Example: It takes 12 tiles to cover this poster.

area model   A visual way of solving multiplication problems by constructing a rectangle with the same dimensions as the numbers you are multiplying and breaking the problem down by place value.

6 × 10 = 60
6 × 8 = 48
so
6 × 18 = 108

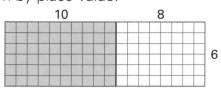

10   8

6

array   An arrangement of items into even columns and rows to make them easier to count.

balance scale   Equipment that balances items of equal mass; used to compare the mass of different items. Also called *pan balance* or *equal arm balance*.

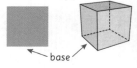

bar graph   A way of representing data using bars or columns to show the values of each variable.

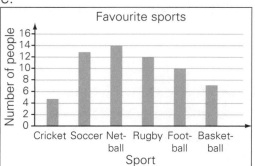

base   The bottom edge of a 2D shape or the bottom face of a 3D shape.

base

capacity   The amount that a container can hold.

Example: The jug has a capacity of 4 cups.

Cartesian plane   A grid system with numbered horizontal and vertical axes that allow for exact locations to be described and found.

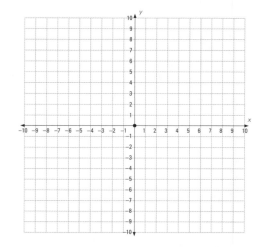

OXFORD UNIVERSITY PRESS

categorical variables   The different groups that objects or data can be sorted into based on common features.

Example: Within the category of ice-cream flavours, variables include:

vanilla   chocolate   strawberry

centimetre or *cm*   A unit for measuring the length of smaller items.

Example: Length is 80 cm.

circle graph   A circular graph divided into sections that look like portions of a pie.

circumference   The distance around the outside of a circle.

clockwise   Moving in the same direction as the hands of a clock.

common denominator   Denominators that are the same. To find a common denominator, you need to identify a multiple that two or more denominators share.

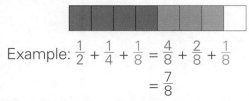

Example: $\frac{1}{2} + \frac{1}{4} + \frac{1}{8} = \frac{4}{8} + \frac{2}{8} + \frac{1}{8}$
$= \frac{7}{8}$

compensation strategy   A way of solving a problem that involves rounding a number to make it easier to work with, and then paying back or "compensating" the same amount.

Example: 24 + 99 = 24 + 100 − 1 = 123

composite number   A number that has more than two factors, that is, a number that is not a prime number.

cone   A 3D shape with a circular base that tapers to a point.

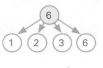

congruent shapes   Shapes that remain the same size and shape even when they have transformed.

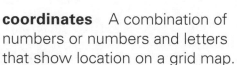

coordinates   A combination of numbers or numbers and letters that show location on a grid map.

corner   The point where two edges of a shape or object meet. Also known as a *vertex*.

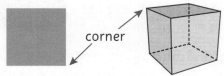

corner

cross-section   The surface or shape that results from making a straight cut through a 3D shape.

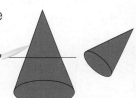

cube   A rectangular prism where all six faces are squares of equal size.

cubic centimetre or *cm³*   A unit for measuring the volume of smaller objects.

Example: This cube is exactly 1 cm long, 1 cm wide and 1 cm deep.

1 cm
1 cm
1 cm

cylinder   A 3D shape with two parallel circular bases and one curved surface.

data   Information gathered through methods such as questioning, surveys or observation.

decimal fraction   A way of writing a number that separates any whole numbers from fractional parts expressed as tenths, hundredths, thousandths and so on.

Example: 1.9 is the same as 1 whole and 9 parts out of 10 or $1\frac{9}{10}$.

degrees Celsius   A unit used to measure the temperature against the Celsius scale where 0°C is the freezing point and 100°C is the boiling point.

denominator   The bottom number in a fraction, which shows how many pieces the whole or group has been divided into.

$\frac{3}{4}$

diameter   A straight line from one side of a circle to the other, passing through the centre point.

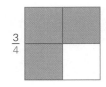

digital time   Time shown on a clock or watch face with numbers only to indicate the hours and minutes.

division/dividing   The process of sharing a number or group into equal parts, with or without remainders.

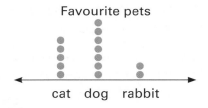

dot plot   A way of representing pieces of data using dots along a line labelled with variables.

Favourite pets

cat   dog   rabbit

double/doubles   Adding two identical numbers or multiplying a number by 2.

Example: $2 + 2 = 4$ $4 \times 2 = 8$

duration   How long something lasts.

Example: Most movies have a duration of about 2 hours.

edge   The side of a shape or the line where two faces of an object meet.

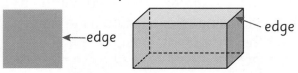

←edge

edge

equal   Having the same number or value.

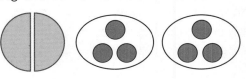

Example: Equal size   Equal numbers

equation   A written mathematical problem where both sides are equal.

Example: $4 + 5 = 6 + 3$

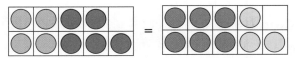

equilateral triangle   A triangle with three sides and angles the same size.

equivalent fractions   Different fractions that represent the same size in relation to a whole or group.

$\frac{1}{2}$ $\frac{2}{4}$ $\frac{3}{6}$ $\frac{4}{8}$

estimate   A thinking guess.

even number   A number that can be divided equally into 2.

Example: 4 and 8 are even numbers

face   The flat surface of a 3D shape.

face →

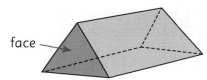

factor   A whole number that will divide evenly into another number.

Example: The factors of 10 are 1 and 10
2 and 5

financial plan   A plan that helps you to organise or manage your money.

flip   To turn a shape over horizontally or vertically. Also known as *reflection*.

horizontal flip

vertical flip

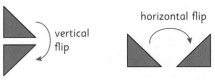

OXFORD UNIVERSITY PRESS

fraction   An equal part of a whole or group.

Example: One out of two parts or $\frac{1}{2}$ is shaded.

grams or _g_   A unit for measuring the mass of smaller items.

1000 g is 1 kg

graph   A visual way to represent data or information.

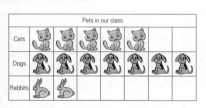

GST or Goods and Services Tax   A tax, such as 10%, that applies to most goods and services bought in many countries.

Example: Cost + GST (10%) = Amount you pay
$10   + $0.10   = $10.10

hexagon   A 2D shape with six sides.

horizontal   Parallel with the horizon or going straight across.

horizontal line

improper fraction   A fraction where the numerator is greater than the denominator, such as $\frac{3}{2}$.

integer   A whole number. Integers can be positive or negative.

inverse operations   Operations that are the opposite or reverse of each other. Addition and subtraction are inverse operations.

Example: 6 + 7 = 13 can be reversed with 13 − 7 = 6

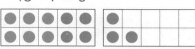

invoice   A written list of goods and services provided, including their cost and any GST.

Priya's Pet Store			
Tax Invoice			
Item	Quantity	Unit price	Cost
Siamese cat	1	$500	$500.00
Cat food	20	$1.50	$30.00
Total price of goods		$530.00	
GST (10%)		$53.00	
Total		$583.00	

isosceles triangle   A triangle with two sides and two angles of the same size.

jump strategy   A way to solve number problems that uses place value to "jump" along a number line by hundreds, tens and ones.

Example: 16 + 22 = 38

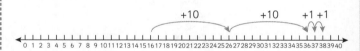

kilograms or _kg_   A unit for measuring the mass of larger items.

kilometres or _km_   A unit for measuring long distances or lengths.

kite   A four-sided shape where two pairs of adjacent sides are the same length.

legend   A key that tells you what the symbols on a map mean.

🌲 Park   ⛽ Service station   🔥 Campground   🚂 Railway | Road

length   The longest dimension of a shape or object.

line graph   A type of graph that joins plotted data with a line.

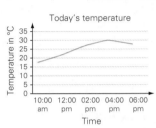

Today's temperature

litres or L   A unit for measuring the capacity of larger containers.

Example: The capacity of this bucket is 8 litres.

mass   How heavy an object is.

Example: 4.5 kilograms   4.5 grams

metre or m   A unit for measuring the length of larger objects.

milligram or mg   A unit for measuring the mass of lighter items or to use when accuracy of measurements is important.

700 mg

millilitre or mL   A unit for measuring the capacity of smaller containers.

1000 mL is 1 litre

millimetre or mm   A unit for measuring the length of very small items or to use when accuracy of measurements is important.

cm   1   2   3

There are 10 mm in 1 cm.

mixed number   A number that contains both a whole number and a fraction.

Example: $2\frac{3}{4}$

multiple   The result of multiplying a particular whole number by another whole number.

Example: 10, 15, 20 and 100 are all multiples of 5.

near doubles   A way to add two nearly identical numbers by using known doubles facts.

Example: 4 + 5 = 4 + 4 + 1 = 9

net   A flat shape that when folded up makes a 3D shape.

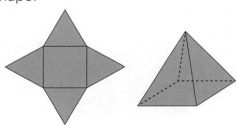

number line   A line on which numbers can be placed to show their order in our number system or to help with calculations.

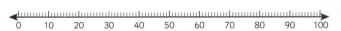

0   10   20   30   40   50   60   70   80   90   100

number sentence   A way to record calculations using numbers and mathematical symbols.

Example: 23 + 7 = 30

OXFORD UNIVERSITY PRESS

numeral   A figure or symbol used to represent a number.

Examples: 1 – one   2 – two   3 – three

numerator   The top number in a fraction, which shows how many pieces you are dealing with.

 $\frac{3}{4}$

obtuse angle   An angle that is larger than a right angle or 90 degrees, but smaller than 180 degrees.

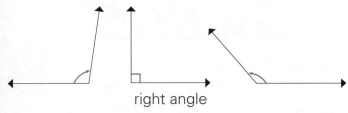

right angle

octagon   A 2D shape with eight sides.

odd number   A number that cannot be divided equally into 2.

Example: 5 and 9 are odd numbers.

operation   A mathematical process. The four basic operations are addition, subtraction, multiplication and division.

origin   The point on a Cartesian plane where the *x*-axis and *y*-axis intersect.

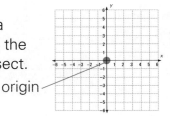

origin

outcome   The result of a chance experiment.

Example: The possible outcomes if you roll a dice are 1, 2, 3, 4, 5 or 6.

parallel lines   Straight lines that are the same distance apart and so will never cross.

parallel   parallel   not parallel

parallelogram   A four-sided shape where each pair of opposite sides is parallel.

pattern   A repeating design or sequence of numbers.

Example:
Shape pattern
Number pattern   2, 4, 6, 8, 10, 12

pentagon   A 2D shape with five sides.

per cent or %   A fraction out of 100.

Example: $\frac{62}{100}$ or 62 out of 100

 is also 62%.

perimeter   The distance around the outside of a shape or area.

Example: Perimeter = 7 m + 5 m + 10 m + 3 m + 6 m = 31 m

pictograph   A way of representing data using pictures so that it is easy to understand.

Example: Favourite juices in our class

place value   The value of a digit depending on its place in a number.

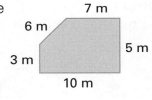

M	H Th	T Th	Th	H	T	O
			2	7	4	8
		2	7	4	8	6
	2	7	4	8	6	3
2	7	4	8	6	3	1

polygon   A closed 2D shape with three or more straight sides.

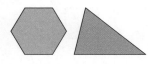

polygons   not polygons

polyhedron (plural polyhedra)   A 3D shape with flat faces.

polyhedra   not polyhedra

power of   The number of times a particular number is multiplied by itself.

Example: 4^3 is 4 to the power of 3 or $4 \times 4 \times 4$.

prime number   A number that has just two factors – 1 and itself. The first four prime numbers are 2, 3, 5 and 7.

prism   A 3D shape with parallel bases of the same shape and rectangular side faces.

triangular   rectangular   hexagonal
prism   prism   prism

probability   The chance or likelihood of a particular event or outcome occurring.

 Example: There is a 1 in 8 chance this spinner will land on red.

protractor   An instrument used to measure the size of angles in degrees.

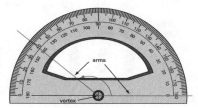

pyramid   A 3D shape with a 2D shape as a base and triangular faces meeting at a point.

square pyramid   hexagonal pyramid

quadrant   A quarter of a circle or one of the four quarters on a Cartesian plane.

quadrant

quadrant

quadrilateral   Any 2D shape with four sides.

radius   The distance from the centre of a circle to its circumference or edge.

reflect   To turn a shape over horizontally or vertically. Also known as *flipping*.

vertical   horizontal reflection
reflection

reflex angle   An angle that is between 180 and 360 degrees in size.

remainder   An amount left over after dividing one number by another.

Example: $11 \div 5 = 2$ r1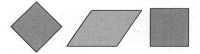

rhombus   A 2D shape with four sides, all of the same length and opposite sides parallel.

right angle   An angle of exactly 90 degrees.

90°

90°

right-angled triangle   A triangle where one angle is exactly 90 degrees.

90°

 rotate   Turn around a point.

OXFORD UNIVERSITY PRESS

rotational symmetry   A shape has rotational symmetry if it fits into its own outline at least once while being turned around a fixed centre point.

1st position

Back to the start

2nd position

round/rounding   To change a number to another number that is close to it to make it easier to work with.

229 can be

rounded up to   OR   rounded down to
the nearest 10   the nearest 100
↑ 230   ↓ 200

scale   A way to represent large areas on maps by using ratios of smaller to larger measurements.

Example: 1 cm = 5 m

scalene triangle   A triangle where no sides are the same length and no angles are equal.

sector   A section of a circle bounded by two radius lines and an arc.

radius lines

arc

sector

semi-circle   Half a circle, bounded by an arc and a diameter line.

semi-circle

arc

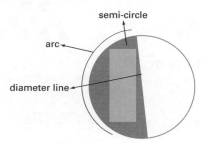

diameter line

similar shapes   Shapes whose angles remain the same size even when the lengths of the sides have been changed.

skip counting   Counting forwards or backwards by the same number each time.

Examples:
Skip counting by fives: 5, 10, 15, 20, 25, 30
Skip counting by twos: 1, 3, 5, 7, 9, 11, 13

slide   To move a shape to a new position without flipping or turning it. Also known as *translate*.

sphere   A 3D shape that is perfectly round.

split strategy   A way to solve number problems that involves splitting numbers up using place value to make them easier to work with.

Example: 21 + 14 =
20 + 10 + 1 + 4 = 35

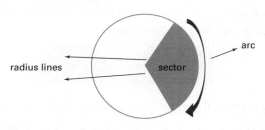

square centimetre or c*m*²
A unit for measuring the area of smaller objects. It is exactly 1 cm long and 1 cm wide.

1 cm
1 cm

square metre or *m*²   A unit for measuring the area of larger spaces. It is exactly 1 m long and 1 m wide.

1 m
1 m

square number   The result of a number being multiplied by itself. The product can be represented as a square array.

Example: 3 × 3 or 3² = 9

straight angle   An angle that is exactly 180 degrees in size.

180°

strategy A way to solve a problem. In mathematics, you can often use more than one strategy to get the right answer.

Example: 32 + 27 = 59
Jump strategy

Split strategy
30 + 2 + 20 + 7 = 30 + 20 + 2 + 7 = 59

subtraction The taking away of one number from another number. Also known as *subtracting, take away, difference between* and *minus*. See also *vertical subtraction*.

Example: 5 take away 2 is 3

survey A way of collecting data or information by asking questions.

Strongly agree	☐
Agree	☑
Disagree	☐
Strongly disagree	☐

symmetry A shape or pattern has symmetry when one side is a mirror image of the other.

table A way to organise information that uses columns and rows.

Flavour	Number of people
Chocolate	12
Vanilla	7
Strawberry	8

tally marks A way of keeping count that uses single lines with every fifth line crossed to make a group.

term A number in a series or pattern.

Example: The sixth term in this pattern is 18.

| 3 | 6 | 9 | 12 | 15 | 18 | 21 | 24 |

tessellation A pattern formed by shapes that fit together without any gaps.

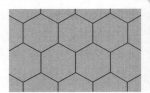

thermometer An instrument for measuring temperature.

three-dimensional or *3D*
A shape that has three dimensions – length, width and depth.
3D shapes are not flat.

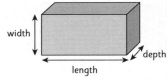

time line A visual representation of a period of time with significant events marked in.

translate To move a shape to a new position without flipping or turning it. Also known as *slide*.

trapezium A 2D shape with four sides and only one set of parallel lines.

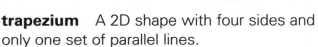

triangular number A number that can be organised into a triangular shape. The first four are:

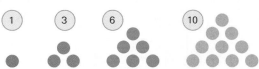

two-dimensional or *2D*
A flat shape that has two dimensions – length and width.

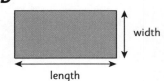

OXFORD UNIVERSITY PRESS

turn   Rotate around a point.

unequal   Not having the same size or value.

Example: Unequal size   Unequal numbers

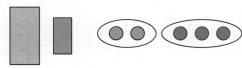

value   How much something is worth.

Example:
This coin is worth 5c.   This coin is worth $1.

vertex (plural vertices)   The point where two edges of a shape or object meet. Also known as a *corner*.

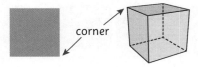

corner

vertical   At a right angle to the horizon or straight up and down.

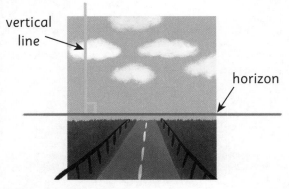

vertical line

horizon

vertical addition   A way of recording addition so that the place-value columns are lined up vertically to make calculation easier.

T	O
3	6
+ 2	1
5	7

vertical subtraction   A way of recording subtraction so that the place-value columns are lined up vertically to make calculation easier.

T	O
5	7
− 2	1
3	6

volume   How much space an object takes up.

Example: This object has a volume of 4 cubes.

whole   All of an item or group.

Example: A whole shape   A whole group

width   The shortest dimension of a shape or object. Also known as *breadth*.

***x*-axis**   The horizontal reference line showing coordinates or values on a graph or map.

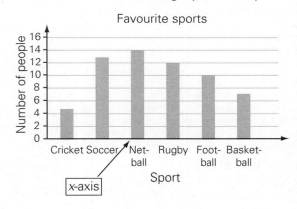

***y*-axis**   The vertical reference line showing coordinates or values on a graph or map.

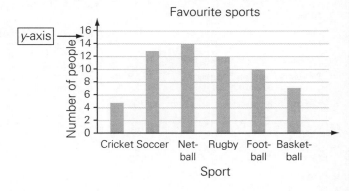

Guided practice

1

	Hundred thousands	Ten thousands	Thousands	Hundreds	Tens	Ones	Write the number using gaps if necessary.
a		2	0	0	0	0	20 000
b			5	0	0	0	5000
c				3	0	0	300
d					8	0	80
e						4	4

2   **a** 9307   **b** 25 046   **c** 102 701

3   **a** two thousand, eight hundred and sixty
    b thirteen thousand, four hundred and sixty-five
    c twenty-eight thousand, seven hundred and five

Independent practice

1   **a** 3000   **b** 8000   **c** 20 000
    d 100 000   **e** 500

2   **a** fifty-three thousand, two hundred and seven
    b forty-eight thousand and five
    c twenty-nine thousand, four hundred and twenty-five
    d one hundred and thirty-five thousand, two hundred and eighty-four
    e three hundred and ninety-nine thousand, five hundred and seventeen

3   **a** 86 231   **b** 142 000
    c 656 308   **d** 105 921

4 25 790

5   **a** 20 000 + 5000 + 100 + 20 + 3
    b 60 000 + 3000 + 300 + 80 + 2
    c 6000 + 4
    d 100 000 + 20 000 + 5000 + 300 + 80 + 1
    e 800 000 + 60 000 + 90 + 4

6   **a** 976 531   **b** 136 795
    c 796 531   **d** 351 679

7   **c** 236 356; two hundred and thirty-six thousand, three hundred and fifty-six
    d 154 009; one hundred and fifty-four thousand and nine

Extended practice

1

Place	Activity	Record number
USA	Number of dogs on a dog walk together	3117
Spain	People salsa dancing together	3 868
Poland	People ringing bells together	10 021
Hong Kong	People playing percussion instruments together	10 102
Singapore	People line dancing together	11 967
Portugal	People making a human advertising sign	34 309
Mexico	People doing aerobics at the same time	38 633
India	Trees planted by a group in one day	80 241
USA	People in a conga line	119 986
England	The longest scarf ever knitted (cm)	322 000

2   **a** 80 241: trees planted
    b 38 633: aerobics
    c 3117: dogs
    d 322 000: scarf
    e 10 102: percussion instruments
    f 10 021: bells
    g 119 986: conga line
    h 11 967: line dancing
    i 3868: salsa dancing
    j 34 309: advertising sign

3 Teacher: Look at the way the student organises the list. The numbers that round to 50 000 must start with either 51 000 or 52 000. Using each of the other 3 digits in turn, the possible numbers are 51 269, 51 296, 51 629, 51 692, 51 926, 51 962, 52 169, 52 196, 52 619, 52 691, 52 916, 52 961. (The actual population was 51 962.)

Guided practice

1

	Problem	Find a near-double	Now I need to:	Answer
e.g.	252 + 250	250 + 250 = 500	add 2 more	502
a	150 + 160	150 + 150 = 300	add 10 more	310
b	126 + 126	125 + 125 = 250	add 2 more	252
c	1400 + 1450	1400 + 1400 = 2800	add 50 more	2850

2

	Problem	Expand the numbers	Join the partners	Answer
e.g.	252 + 250	200 + 50 + 2 + 200 + 50	200 + 200 + 50 + 50 + 2 = 500 + 2	502
a	66 + 34	60 + 6 + 30 + 4	60 + 30 + 6 + 4 = 90 + 10	100
b	140 + 230	100 + 40 + 200 + 30	100 + 200 + 40 + 30 = 300 + 70	370
c	1250 + 2347	1000 + 200 + 50 + 2000 + 300 + 40 + 7	1000 + 2000 + 200 + 300 + 50 + 40 + 7	3597

3   **a** What is 105 + 84?   **b** What is 1158 + 130?   **c** What is 2424 + 505?

    Answer: 105 + 84 = 189   Answer: 1158 + 130 = 1288   Answer: 2424 + 505 = 2929

Independent practice

1

	Problem	Using rounding it becomes:	Now I need to:	Answer
a	56 + 41	56 + 40 = 96	add 1	97
b	25 + 69	25 + 70 = 95	take away 1	94
c	125 + 62	125 + 60 = 185	add 2	187
d	136 + 198	136 + 200 = 336	take away 2	334
e	195 + 249	195 + 250 = 445	take away 1	444
f	1238 + 501	1238 + 500 = 1738	add 1	1739
g	1645 + 1998	1645 + 2000 = 3643	take away 2	3643

2 Student may choose a different strategy to the one suggested. Teachers may wish to ask students to explain (perhaps to the group) how they arrived at one or two of the answers.
    a 134   **b** 125   **c** 371
    d 2409   **e** 2950   **f** 2566

3 Students may choose a different strategy to the one suggested. Teachers may wish to ask students to explain (perhaps to the group) how they arrived at one or two of the answers.
    a 163   **b** 211
    c 2035   **d** 3906

4

	Problem	Expand the numbers	Join the partners	Answer
e.g.	125 + 132	100 + 20 + 5 + 100 + 30 + 2	100 + 100 + 20 + 30 + 5 + 2	257
a	173 + 125	100 + 70 + 3 + 100 + 20 + 5	100 + 100 + 70 + 20 + 3 + 5	298
b	1240 + 2130	1000 + 200 + 40 + 2000 + 100 + 30	1000 + 2000 + 200 + 100 + 40 + 30	3370
c	5125 + 1234	5000 + 100 + 20 + 5 + 1000 + 200 + 30 + 4	5000 + 1000 + 100 + 200 + 20 + 30 + 5 + 4	6359
d	7114 + 2365	7000 + 100 + 10 + 4 + 2000 + 300 + 60 + 5	7000 + 2000 + 100 + 300 + 10 + 60 + 4 + 5	9479
e	2564 + 4236	2000 + 500 + 60 + 4 + 4000 + 200 + 30 + 6	2000 + 4000 + 500 + 200 + 60 + 30 + 4 + 6	6800

5 Teachers may wish to ask students to explain (perhaps to the group) how they arrived at some of the answers.
    a 903   **b** 2980   **c** 6027
    d 4998   **e** 3501   **f** 1483
    g 4998   **h** 5490

Extended practice

1 **a** 2200   **b** 1500   **c** 4800
   d 4500   **e** 8900   **f** 2200
   g 600 000   **h** 200 000

2 **a** 3700 m   **b** 300 m
   c 800 km (800 000 m)

3 **a** $1
   b The ball (99c rounds to $1)

UNIT 1: Topic 3

Guided practice

1 **a** 49   **b** 274
   c 498   **d** 4866

2 **a** 86   **b** 284
   c 425   **d** 917

3 **a** 386   **b** 4623
   c 47 823   **d** 75 120   **e** 700 131

Independent practice

1 **a** 123   **b** 1234   **c** 12 345
   d 123 456   **e** 121   **f** 2332
   g 34 543   **h** 456 654   **i** 111
   j 2222   **k** 33 333   **l** 444 444

2 **a** 90   **b** 820   **c** 815
   d 1320   **e** 2307

3 **a** No. (Teachers may ask students to justify their response, e.g the answer is not a reasonable one because $300 + $200 + $1000 + $100 + $200 = $1800
   b $1792

4 **a** 251   **b** 1065   **c** 1017
   d 244   **e** 1140   **f** 1543
   g 4027   **h** 38 373   **i** 62 070
   j 12 257

Extended practice

1 There are two possible answers: 335 or 435. Look for students who solve the problem systematically.
The realistic addends for 335 are 319 + 16 and 309 + 26 although 329 + 06 will give the same answer.
Addends for 435 are:
399 + 36   389 + 46
379 + 56   369 + 66
359 + 76   349 + 86
339 + 96

2 Multiple answers are possible. Teachers may wish to ask students to use a calculator to check the total. An easy solution would be to subtract 1 from the average for the first game and add 1 to the average for the second. Then subtract 2 from the average for the third game and add 2 for the fourth, and so on.

3 The answer is 123 456.

UNIT 1: Topic 4

Guided practice

1

	Problem	Using rounding it becomes	Now I need to:	Answer
a	53 – 21	53 – 20 = 33	take away 1	32
b	85 – 28	85 – 30 = 55	add 2	57
c	167 – 22	167 – 20 = 147	take away 2	145
d	146 – 198	346 – 200 = 146	add 2	148
e	1787 – 390	1787 – 400 = 1387	add 10	1397
f	5840 – 3100	5840 – 3000 = 2840	take away 100	2740
g	6178 – 3995	6178 – 4000 = 2178	add 5	2183

2

	Problem	Expand the number	Take away 1st part	Take away 2nd part	Take away 3rd part	Answer
a	257 – 126	126 = 100 + 20 + 6	257 – 100 = 157	157 – 20 = 137	137 – 6 = 131	131
b	548 – 224	224 = 200 + 20 + 4	548 – 200 = 348	348 – 20 = 328	328 – 4 = 324	324
c	765 – 442	442 = 400 + 40 + 2	765 – 400 = 365	365 – 40 = 325	325 – 2 = 323	323
d	878 – 236	236 = 200 + 30 + 6	878 – 200 = 678	678 – 30 = 648	648 – 6 = 42	642
e	999 – 753	753 = 700 + 50 + 3	999 – 700 = 299	299 – 50 = 749	249 – 3 = 746	246

Independent practice

1 Students may choose a different strategy from the one suggested. Teachers may wish to ask students to explain (perhaps to the group) how they arrived at one or two of the answers.
   a 25   **b** 155   **c** 316
   d 1236   **e** 3246

2 Students may choose a different strategy from the one suggested. Teachers may wish to ask students to explain (perhaps to the group) how they arrived at one or two of the answers.
   a 21   **b** 121   **c** 422
   d 2402   **e** 3323

3 **a** What is 776 – 423?
       −3 −20   −400
                      776
      353 356 376
      Answer: 776 – 423 = 353

   b What is 487 – 264?
       −4 −60   −200
                      487
      223 227 287
      Answer: 487 – 264 = 223

   c What is 1659 – 536?
       −6 −30   −500
                    1659
      1123 1129 1159
      Answer: 1659 – 536 = 1123

4 Teachers may wish to ask students to explain (perhaps to the group) how they arrived at one or two of the answers.
   a $2.50   **b** $1.25   **c** $6.50
   d $5.55   **e** $4.65   **f** $7.85

5 Teachers may wish to ask students to explain (perhaps to the group) how they arrived at one or two of the answers.
   a 43   **b** 22   **c** 65
   d 33   **e** 115   **f** 110

6 Teachers may wish to ask students to explain (perhaps to the group) how they arrived at one or two of the answers.
   a 70   **b** 51   **c** 57
   d 75   **e** 295   **f** 550

Extended practice

1 1 hour 35 minutes or 95 minutes.

2 Answers will vary. Teachers may wish to ask students to explain (perhaps to the group) how they arrived at their answers. One simple solution is to start with a round number, say 100 and the other number is then 157. The other solutions could then be arrived at by adding 1 to each number (101 and 158, 102 and 159, etc.).

3 Answers will vary. A simple solution is to count up to $5 from $2.45 and the $2.55 then becomes the price of the item.

4 3838
Look for students applying the process of rounding. A simple strategy is to round 397 to 400. 4235 – 400 = 3835. 3 are added back to the number, giving an answer of 3838

5 Bill: $7657, Bob: $7850

6 Teacher to check, e.g. 623 – 545 = 78, 633 – 555 = 78 and 643 – 565 = 78. Look for students who see the pattern of increasing each of the tens by one.

UNIT 1: Topic 5

Guided practice

1 **a** 49   **b** 116   **c** 219
   d 407   **e** 6126   **f** 3094
   g 1506   **h** 3998   **i** 22 187
   j 18 529   **k** 33 247   **l** 567 639

Independent practice

1 a 321   **b** 432   **c** 543
   d 654   **e** 765

2 a 1234   **b** 2345   **c** 3456
   d 4567   **e** 5678   **f** 6789
   g 9876   **h** 8765

3 a 11 111   **b** 22 222   **c** 33 333
   d 44 444   **e** 55 555   **f** 66 666

4 764 321 − 123 467 = 640 854

5 724

6 a First option: 124
   b Second option: 6194
   c First option: 7258

7 a 268   **b** 258   **c** 425
   d 148   **e** 369   **f** 818
   g 13 677   **h** 385 926

Extended practice

1 Multiple answers are possible. Look for students who understand that the lowest 5-digit numbers that have a difference of 999 must be around 11 000 and 10 000. The lowest three possibilities are:

10 999 − 10 000; 11 000 − 10 001; 11 001 − 10 002

2 a 36 831   **b** 11 812
   c 56 149   **d** 25 000

3 978 mm

UNIT 1: Topic 6

Guided practice

1

	a		**b**		**c**		**d**		**e**		**f**	
	t	o	t	o	t	o	t	o	h	t	o	h	t	o
×		7		8		6		9	1	4		1	9
10	7	0	8	0	6	0	9	0	1	4	0	1	9	0

2 a 15 m   **b** 22 L
   c 45 t   **d** $17 (or $17.00)
   e 38 cm   **f** 36 m
   g $27.50 (or $27.5)

3 a 1400   **b** 1700   **c** 1300
   d 2700   **e** 2300   **f** 4500
   g 6400   **h** 370   **i** $125

Independent practice

1 a 6 × 3 tens = 18 tens; 18 tens = 180
   b 9 × 2 tens = 18 tens = 180.
     9 × 3 tens = 27 tens = 270
   c 8 × 2 tens = 16 tens = 160.
     8 × 3 tens = 24 tens = 240
   d 7 × 2 tens = 14 tens = 140.
     7 × 3 tens = 21 tens = 210

2 a 10, 20, 40   **b** 24, 48, 96
   c 30, 60, 120   **d** 100, 200, 400
   e 80, 160, 320

3

	Problem	Double and halve	Product
a	3 × 14	6 × 7	42
b	5 × 18	10 × 9	90
c	3 × 16	6 × 8	48
d	5 × 22	10 × 11	110
e	6 × 16	12 × 8	96
f	4 × 18	8 × 9	72

4

×5	First multiply by 10	Then halve it	Multiplication fact
a	16	160	80	16 × 5 = 80
b	18	180	90	18 × 5 = 90
c	24	240	120	24 × 5 = 120
d	32	320	160	32 × 5 = 160
e	48	480	240	48 × 5 = 240

5 Teachers may wish to ask students to explain (perhaps to the group) how they arrived at one or two of the answers.
   a 180   **b** 1400   **c** 25 m
   d 340   **e** 280   **f** 750
   g 104   **h** 360   **i** $17.50
   j 480

Extended practice

1

	× 15	× 10	Halve it to find × 5	Add the two answers	Multiplication fact
e.g.	12	120	60	120 + 60 = 180	12 × 15 = 180
a	16	160	80	160 + 80 = 240	16 × 15 = 240
b	14	140	70	140 + 70 = 210	14 × 15 = 210
c	20	200	100	200 + 100 = 300	20 × 15 = 300
d	30	300	150	300 + 150 = 450	30 × 15 = 450
e	25	250	125	250 + 125 = 375	25 × 15 = 375

UNIT 1: Topic 7

Guided practice

1 7 × 34 = 7 × 30 + 7 × 4
      = 210 + 28
      = 238

7 × 30 = 210   7 × 4 = 28

2 5 × 28 = 5 × 20 + 5 × 8
      = 100 + 40
      = 140

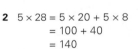

5 × 20 = 100   5 × 8 = 40

Independent practice

1 6 × 32 = 6 × 30 + 6 × 2
      = 180 + 12
      = 192

2 5 × 35 = 5 × 30 + 5 × 5
      = 150 + 25
      = 175

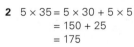

3 7 × 48 = 7 × 40 + 7 × 8
      = 280 + 56
      = 336

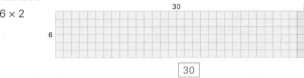

Guided practice

1 a 172   **b** 195   **c** 58
   d 644   **e** 152

2 a 250   **b** 568   **c** 759
   d 975   **e** 2490
   f 696   **g** 1425   **h** 6492
   i 6360   **j** 8692   **k** 9856

Independent practice

1 a 6492   **b** 6936   **c** 7548
   d 21 150   **e** 36 978   **f** 43 076
   g 235 480   **h** 119 260   **i** 181 870
   j 222 633

---

2 Choice 2 is the better choice. Because of doubling, the 4-weekly amounts are 40c + 80c + $1.60 + $3.20 + $6.40 + $12.80 + $25.60 + $51.20 + $102.40 + $204.80 + $409.60 + $819.20, making a total for the year of $1638.

3 The total number of pages is 330. Look for students who use time-saving strategies. For example, you multiply 48 × 5; possible strategy: 48 × 10 = 480. Half of 480 = 240. Then you double 45 = 90. 240 + 90 = 330

2 a 111 111   **b** 222 222   **c** 333 333
   d 444 444   **e** 555 555   **f** 666 666
   g 777 777   **h** 888 888   **i** 999 999

3 a $81.75   **b** $93.75   **c** $87.30

4 a 340   **b** 280   **c** 480
   d 640   **e** 810

5 a 360   **b** 368   **c** 475
   d 624   **e** 555   **f** 855

Extended practice

1 a 29 238 km   **b** 44 178 km
   c 184 944 km **d** 176 008 km
   e Yes, 100 × 10 000 = 1 million.
     (Exact answer = 1 032 600 km)

OXFORD UNIVERSITY PRESS

2 a 13 020 points
   b 152 640 points
   c 130 464 points
3 There is more than one strategy that students could use to solve the problem. Teachers may wish to ask students to discuss how they intend to solve it. Students may opt to double the distance (651 × 2) to find the length of a return journey and then multiply 1302 by 14. Others may choose to multiply 651 by 14 and double the answer.
A third strategy could be to multiply 651 by 7 days, doubling the answer because there are two trips per day and finally doubling again for the return trips.
The total distance is 18 228 km.

UNIT 1: Topic 8

Guided practice

1 a 1 ,2, 4, 8   **b**   1, 5
   c 1, 3, 9   **d** 1, 2, 3, 6
   e 1, 2   **f** 1, 2, 4
   g 1, 7   **h** 1, 3
2 a 3, 6, 9, 12, 15, 18, 21, 24, 27, 30
   b 6, 12, 18, 24, 30, 36, 42, 48, 54, 60
   c 9, 18, 27, 36, 45, 54, 63, 72, 81, 90
   d 2, 4, 6, 8, 10, 12, 14, 16, 18, 20
   e 4, 8, 12, 16, 20, 24, 28, 32, 36, 40
   f 8, 16, 24, 32, 40, 48, 56, 64, 72, 80
   g 7, 14, 21, 28, 35, 42, 49, 56, 63, 70
   h 5, 10, 15, 20, 25, 30, 35, 40, 45, 50

Independent practice

1 a 1, 3, 5, 15   **b**   1, 2, 4, 8, 16
   c 1, 2, 4, 5, 10, 20   **d** 1, 13
   e 1, 2, 7, 14   **f**   1, 2, 3, 6, 9, 18
2 a 23 (1 & 23), 29 (1 & 29)
   b 21 (1, 3, 7, 21), 22 (1, 2, 11, 22), 26 (1, 2, 13, 26), 27 (1, 3, 9, 27)
   c 25 (1, 5, 25)
   d 28 (1, 2, 4, 7, 14, 28)
3 a 1, 2, 3, 4, 6, 8, 12, 24
   b 36 (1, 2, 3, 4, 6, 9, 12, 18, 36)
4 a 4: 1, 2, 4; 8: 1, 2, 4, 8; common factors are 1, 2 and 4
   b 6: 1, 2, 3, 6; 8: 1, 2, 4, 8; common factors are 1 and 2
   c 14: 1, 2, 7, 14; 21: 1, 3, 7, 21; common actors are 1 and 7
   d 12: 1, 2, 3, 4, 6, 12; 18: 1, 2, 3, 6, 9, 18; common factors are 1, 2, 3 and 6
5 a 15, 25, 40, 50, 60, 65, 75, 85, 100
   b 8, 12, 24, 28, 36, 40, 48
   c 8, 16, 24, 32, 48, 56
   d 14, 21, 28, 35, 42, 49, 56
   e 9, 18, 27, 36, 45, 63, 72
6 Teacher to check, e.g.
   a … because it is an even number
   b … because the sum of the digits is divisible by 3
   c … because it does not end in a zero
   d … because all multiples of 5 end in zero or 5

7 2: 2, 4, 6, 8, 10, 12, 14, 16, 18, 20, 22, 24, 26, 28, 30
   3: 3, 6, 9, 12, 15, 18, 21, 24, 27, 30
   Common multiples are 6, 12, 18, 24 and 30
8 20
9 36
10 a 18   **b** 12   **c** 35   **d** 15   **e** 45   **f** 28

Extended practice

1 a 1, 2, 5, 10, 25
   b Possible answers include 5, 10 or 25. Look for students who are able to offer sensible justification for their answers, such as making a packet size that is easily shared by different numbers of people.
2 a 16, 24, 36, 52, 96   **b** 240
   c 24, 30, 36, 90, 96   **d** 24, 36, 96
3 1, 2, 3, 4, 6, 8, 12, 24, 32, 48, 96
4 a 9: Packs of 1, 2, 4, 5, 10, 20, 25, 50 or 100
   b 2, 4, 10, 20, 50 or 100

UNIT 1: Topic 9

Guided practice

1 18, 78, 514, 1000, 1234, 990 and 118
2 a No (e.g. 2, 6, 10, etc.)
   b 4, 8, 12, 16, 20, 24, 28, 32, 36, 40 (Students should recognise multiples of 4 as 4 times table)
3 Teacher to check, e.g. the two digits make a number that is a multiple of 4.
4 112, 620, 428, 340, 716, 412
5 Teacher to check. Look for students who are able to apply the learning about divisibility by 2 and 4 to identify numbers that meet the criteria in the given range.

Independent practice

1 a 411, 207, 513   **b** 775, 630
   c 702, 522   **d** 888, 248
   e 819, 693, 252   **f** 820, 990
2 31
3 a 32, 36, 40   **b** 36   **c** 32, 40
   d 36   **e** 32, 40
   f 33   **g** 36
4 1, 13 and 39
5 The sum of its digits is divisible by 3.
6 9324
7 a Teacher to check, e.g. The last number in the last two digits is 46 and you can't make groups of 4 out of 46 so the whole number is not divisible by 4.
   b Yes
   c Because the sum of the digits (12) is divisible by 3.
   d 2

Extended practice

1 All of them are divisible by 6, 2 and 3.
2 Divisible only by 3: 15, 45, 81
   Divisible only by 4: 20, 44, 76, 92
   Divisible by both: 48, 72, 96
3 a Teacher to check, e.g. Because it is an even number and the sum of the digits is divisible by 3.
   b 1, 2, 3, 6 and 9

4 720
5 Teacher to check. Look for students who can correctly place multiples of 4 in the left oval, multiples of 5 in the right oval and that the overlapping area contains multiples of 20.

UNIT 1: Topic 10

Guided practice

1 a 68 ÷ 2 is the same as 60 ÷ 2 and 8 ÷ 2
   60 ÷ 2 = 30
   8 ÷ 2 = 4
   So 68 ÷ 2 = 30 + 4 = 34
   b 69 ÷ 3 is the same as 60 ÷ 3 and 9 ÷ 3
   60 ÷ 3 = 20
   9 ÷ 3 = 3
   So 69 ÷ 3 = 20 + 3 = 23
   c 84 ÷ 2 is the same as 80 ÷ 2 and 4 ÷ 2
   80 ÷ 2 = 40
   4 ÷ 2 = 2
   So 84 ÷ 2 = 40 + 2 = 42
   d 124 ÷ 4 is the same as 100 ÷ 4 and 24 ÷ 4
   100 ÷ 4 = 25
   24 ÷ 4 = 6
   So 124 ÷ 4 = 25 + 6 = 31
   e 122 ÷ 2 is the same as 100 ÷ 2 and 22 ÷ 2
   100 ÷ 2 = 50
   22 ÷ 2 = 11
   So 122 ÷ 2 = 50 + 11 = 61
   f 145 ÷ 5 is the same as 100 ÷ 5 and 45 ÷ 5
   100 ÷ 5 = 20
   45 ÷ 5 = 9
   So 145 ÷ 5 = 20 + 9 = 29

Independent practice

Note: some students may choose to bypass this written strategy and solve some or all of the problems using mental strategies.

1 a 14   **b** 18   **c** 17   **d** 13
   e 24   **f** 12   **g** 19   **h** 19
   i 14   **j** 29   **k** 12   **l** 13
2 a 117   **b** 112   **c** 217   **d** 425
   e 116   **f** 318   **g** 117   **h** 114
   i 337   **j** 115   **k** 215   **l** 224
   m 126   **n** 113   **o** 449   **p** 114
3 a 87   **b** 54   **c** 48   **d** 34
   e 22   **f** 67   **g** 54   **h** 57
   i 52   **j** 47   **k** 85   **l** 93
   m 98   **n** 79   **o** 92   **p** 99
4 a 14 r1   **b** 25 r1   **c** 15 r2
   d 13 r2   **e** 115 r3   **f** 317 r1
   g 116 r2   **h** 111 r5   **i** 55 r2
   j 45 r5   **k** 66 r1   **l** 55 r2
   m 41 r6   **n** 43 r7   **o** 68 r3
   p 99 r1

Extended practice

1 a 19 r2   **b** 24
   c 24 r1   **d** 55 r1

2 Students' own answers. Look for students who use remainders appropriately and who recognise that donuts can be easily split (whereas marbles cannot) and that dollars can be divided into dollars and cents.
  a $3\frac{1}{2}$ each
  b 4 marbles each and one is left over
  c $6.50

3 a The average is 161 ÷ 6 = 26 r5 or 26.83. Students may opt to round up the number, and this could be a useful discussion point.
  b Look for the strategies that the students choose to solve the problem. Having found the average number per class to be around 26, students could subtract the total of the numbers shown from the number in the six classes (161 – 51). The total of the four remaining classes should therefore be 110. Appropriate class sizes might be 24 + 27 + 29 + 30, but there are other possibilities.

4 a $33.33 (Students may choose to round the figure to $33.35 but this should lead to reflection that the total would need to be $100.05 for each person to receive that amount. A simpler solution might be to take $33.30 each and put 10c in a charity box!)
  b Depending on the way the student splits the $100, an appropriate way of having $33.30 could be 1 × $20, 1 × $10. 1 × $2, 1 × $1, 1 × 20c and 1 × 10c

5 32 (3000 ÷ 96 = 31 r1, 31.25 or $31\frac{1}{4}$, so 32 boxes are needed)

UNIT 2: Topic 1

Note for teachers: In answering some of the questions, students could choose to write fractions of equivalent value, e.g. $\frac{1}{2}$ instead of $\frac{3}{6}$.

Guided practice

1 a $\frac{1}{6}$ **b** one fifth, $\frac{1}{5}$
  c one third, $\frac{1}{3}$ **d** one eighth, $\frac{1}{8}$
2 Student shades:
  a 3 parts   **b** 3 parts
  c 2 parts   **d** 3 parts
  e 5 parts
3 a $\frac{2}{5}$ **b** $\frac{1}{6}$ **c** $\frac{5}{8}$ **d** $\frac{7}{10}$
4 Student shades:
  a 3 triangles   **b** 5 circles
  c 2 stars   **d** 4 hexagons

Independent practice

1 a
  b
  c
  d
  e
  f
  g

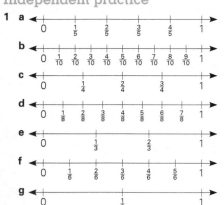

---

2 a $\frac{3}{8}$ **b** $\frac{1}{3}$ **c** $\frac{1}{4}$
  d $\frac{1}{5}$ **e** $\frac{1}{2}$ **f** $\frac{7}{8}$
  g $\frac{4}{5}$ **h** $\frac{5}{8}$ **i** $\frac{7}{8}$
3 $\frac{5}{10}, \frac{2}{4}, \frac{4}{8}, \frac{3}{6}$
4 a $\frac{1}{5}, \frac{2}{5}, \frac{3}{5}, \frac{4}{5}, 1$ **b** $\frac{2}{10}, \frac{3}{10}, \frac{6}{10}, \frac{7}{10}, \frac{9}{10}, 1$
  c $\frac{1}{10}, \frac{1}{8}, \frac{1}{5}, \frac{1}{4}, \frac{1}{2}$ **d** $\frac{3}{10}, \frac{3}{8}, \frac{3}{6}, \frac{3}{4}, \frac{3}{3}$
  e $\frac{2}{10}, \frac{2}{8}, \frac{2}{6}, \frac{2}{5}, \frac{2}{3}$
5 a $\frac{3}{4} < \frac{7}{8}$ **b** $\frac{1}{4} > \frac{1}{8}$ **c** $\frac{3}{6} = \frac{1}{2}$
  d $\frac{2}{3} > \frac{2}{6}$ **e** $\frac{3}{8} < \frac{1}{2}$ **f** $\frac{2}{4} < \frac{5}{8}$
  g $\frac{9}{10} > \frac{4}{5}$ **h** $\frac{3}{5} = \frac{6}{10}$ **i** $\frac{5}{6} > \frac{2}{3}$
6 a $\frac{6}{8}$ and $\frac{3}{4}$ **b** $\frac{2}{8}$
  c Student draws a diamond at $\frac{3}{8}$.
7 Teacher to check and to decide on level of accuracy.
  a Student should attempt to split the rectangle into 8 approximately equal parts.
  b Student shades two parts.
  c $\frac{1}{4}$ (or any equivalent fraction).

Extended practice

1 a Students should see that the guide marks will split the rectangle into twelfths and divide the rectangle at the 4th and 8th marks.
  b Students shade $\frac{1}{3}$ of the rectangle.
  c $\frac{1}{3}$ and $\frac{4}{12}$ (or equivalent fractions).
2
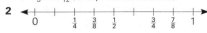
3 Students' own answers. Look for students who demonstrate understanding of fraction sizes by accurately selecting fractions that meet the criteria given.
4 Teacher to check.
  It is unlikely that the student will be able to fold the paper more than six times.
  One fold will divide the paper into halves.
  Two folds will divide the paper into $\frac{1}{4}$s.
  Three folds will divide the paper into $\frac{1}{8}$s.
  Four folds will divide the paper into $\frac{1}{16}$s.
  Five folds will divide the paper into $\frac{1}{32}$s.
  Six folds will divide the paper into $\frac{1}{64}$s.
  Seven folds will divide the paper into $\frac{1}{128}$s.
  Eight folds will divide the paper into $\frac{1}{256}$s.

UNIT 2: Topic 2

Guided practice

Teacher: Allow for equivalent fractions in any or all answers.

1 a 2 quarters; $\frac{2}{4}$ **b** 3 eighths; $\frac{1}{8} + \frac{2}{8} = \frac{3}{8}$
  c 4 fifths; $\frac{2}{5} + \frac{2}{5} = \frac{4}{5}$ **d** 5 sixths; $\frac{2}{6} + \frac{3}{6} = \frac{5}{6}$
  e $\frac{2}{4}$ **f** $\frac{2}{3} - \frac{1}{3} = \frac{1}{3}$

Independent practice

1 a $\frac{3}{8} + \frac{2}{8} = \frac{5}{8}$ **b** $\frac{2}{5} + \frac{1}{5} = \frac{3}{5}$
  c $\frac{2}{6} + \frac{1}{6} = \frac{3}{6}$ **d** $\frac{2}{4} - \frac{1}{4} = \frac{1}{4}$
  e $\frac{3}{3} - \frac{1}{3} = \frac{2}{3}$

---

2 Teacher to check shading
  a $\frac{4}{5}$ **b** $\frac{4}{6}$ **c** $\frac{7}{8}$
  d $\frac{3}{3}$ (or 1 whole)   **e** $\frac{7}{10}$
3 a $\frac{3}{8}$ **b** $\frac{6}{10}$ **c** $\frac{1}{6}$
  d $\frac{2}{5}$ **e** $\frac{1}{3}$
4 a $\frac{9}{8}$ or $1\frac{1}{8}$ **b** $\frac{8}{6}$ or $1\frac{2}{6}$
5 a $\frac{3}{4} + \frac{2}{4} = \frac{5}{4} = 1\frac{1}{4}$ **b** $1\frac{3}{8} - \frac{4}{8} = \frac{7}{8}$
6 a $\frac{6}{4}$ or $1\frac{2}{4}$ **b** $\frac{6}{8}$
  c $\frac{7}{5}$ or $1\frac{2}{5}$ **d** $\frac{4}{6}$
  e $\frac{13}{10}$ or $1\frac{3}{10}$ **f** $\frac{2}{3}$

Extended practice

1 Teacher to check shading
  a $\frac{1}{6} + \frac{3}{6} = \frac{4}{6}$ **b** $\frac{4}{10} + \frac{1}{5} = \frac{4}{10} + \frac{2}{10} = \frac{6}{10}$
2 a $\frac{5}{10}$ (or equivalent)   **b** $\frac{4}{6}$ (or equivalent)
  c $\frac{3}{4}$ **d** $\frac{9}{10}$
  e $\frac{1}{4}$ **f** $\frac{9}{8}$ or $1\frac{1}{8}$
  g $\frac{6}{6}$ or 1 whole   **h** $\frac{7}{8}$

UNIT 2: Topic 3

Guided practice

1 a $\frac{2}{100}$, 0.02
  b 70 tenths, $\frac{7}{10}$, 0.7
  c 9 hundredths, $\frac{9}{100}$, 0.09
  d 26 hundredths, $\frac{26}{100}$, 0.26
  e 89 hundredths, $\frac{89}{100}$, 0.89
2 Student shades as follows:
  a any 40 squares   **b** any 4 squares
  c any 15 squares   **d** any 70 squares
  e 99 squares
3 a 0.3   **b** 0.23   **c** 0.03
4 a $\frac{6}{10}$ **b** $\frac{77}{100}$ **c** $\frac{8}{100}$

Independent practice

1 a 0.004 $\frac{4}{1000}$
  b 0.13 $\frac{13}{1000}$
  c 0.124 $\frac{124}{1000}$
2 a 0.125   **b** 0.008   **c** 0.087
  d 0.002   **d** 0.022   **e** 0.099
3 a $\frac{5}{1000}$ **b** $\frac{255}{1000}$ **c** $\frac{101}{1000}$
  d $\frac{35}{1000}$ **e** $\frac{999}{1000}$ **f** $\frac{9}{1000}$
4 $\frac{14}{627}$
5 a 0.01 > 0.001   **b** $\frac{3}{1000}$ = 0.003
  c $\frac{25}{1000}$ < 0.25   **d** 0.003 < 0.2
  e $\frac{125}{1000}$ = 0.125   **f** $\frac{6}{1000}$ < 0.01
  g 0.02 > $\frac{2}{1000}$ **h** 1 > 0.999
  i $\frac{19}{1000}$ < 0.19   **j** 0.052 = $\frac{52}{1000}$
  k 0.430 > 0.043   **l** 0.999 = $\frac{999}{1000}$
6 a

| 0 | 0.1 | 0.2 | 0.3 | 0.4 | 0.5 | 0.6 | 0.7 | 0.8 | 0.9 | 1 |

b

| 0 | 0.01 | 0.02 | 0.03 | 0.04 | 0.05 | 0.06 | 0.07 | 0.08 | 0.09 | 0.1 |

c

| 0 | 0.001 | 0.002 | 0.003 | 0.004 | 0.005 | 0.006 | 0.007 | 0.008 | 0.009 | 0.01 |

OXFORD UNIVERSITY PRESS

7 a 0.1, 0.2, 0.4, 0.5, 0.9
b 0.02, 0,03. 0.04, 0.06, 0.07
c 0.001, 0.002, 0.004, 0.007, 0.008
d 0.002, 0.02, 0.1, 0.2, 0.3
e 0.1, 0.11, 0.15, 0.2, 0.22
f 0.005, 0.05, 0.055, 0.5, 0.555

Extended practice

1 a 0.1 (Accept 0.10. This could prove an interesting discussion point, particularly when decimals are used with money.)
b 0.045

2 $0.05

3 a $0.25   **b** $0.08
c $0.15   **d** $0.75
e $0.20 (Accept $0.2. This could prove an interesting discussion point when students complete question 4.)
f $0.80   **g** $1.15   **h** $2.20

4 $2.9 \times 3 =$

5 a $7.90   **b** $8.10   **c** $13.20
d $5.75   **e** $13.85

UNIT 2: Topic 4

Note: Teacher to decide the extent to which equivalent fractions, such as $\frac{1}{10}$ for $\frac{10}{100}$, are expected in this topic.

Guided practice

1 a 0.03, 3%   **b** $\frac{9}{100}$, 0.09, 9%
c $\frac{1}{10}$, 0.1, 10%   **d** $\frac{3}{10}$, 0.3, 30%
e $\frac{95}{100}$, 0.95, 95%   **f** $\frac{99}{100}$, 0.99, 99%

2 a $\frac{20}{100}$, 0.2, 20%
Student shades any 20 squares.
b $\frac{15}{100}$, 0.15, 15%
Student shades any 15 squares.
c $\frac{75}{100}$, 0.75, 75%
Student shades any 75 squares.
d $\frac{55}{100}$, 0.55, 55%
Student shades any 55 squares.

Independent practice

1

2

	Fraction	Decimal	Percentage
a	$\frac{5}{100}$	0.05	5%
b	$\frac{25}{100}$	0.25	25%
c	$\frac{75}{100}$	0.75	75%
d	$\frac{99}{100}$	0.99	99%
e	$\frac{9}{10}$	0.9	90%
f	$\frac{4}{10}$	0.4	40%
g	$\frac{1}{10}$	0.1	10%
h	$\frac{2}{100}$	0.02	2%
i	$\frac{3}{10}$	0.3	30%
j	1	1	100%
k	$\frac{1}{2}$	0.5	50%
l	$\frac{1}{100}$	0.01	1%

3 a true   **b** false   **c** false
d true   **e** true   **f** true
g false   **h** true   **i** false

4 a Student shades 50 squares.
$\frac{1}{2}$ is the same as 50%
b Student shades 25 squares.
$\frac{1}{4}$ is the same as 25%
c Student shades 75 squares.
$\frac{3}{4}$ is the same as 75%

5 a $\frac{2}{100}$, 0.03, 20%
b 0.05, 6%, 0.5
c 5%, $\frac{1}{2}$, $\frac{55}{100}$
d 0.04, $\frac{1}{4}$, 40%
e 0.07, 70%, $\frac{3}{4}$
f 0.01. 10%, $\frac{11}{100}$

6 Student colours 3 circles red, 4 circles blue and 3 circles yellow.

7 $\frac{7}{10}$, 0.7, 70%

8 Student colours 4 diamonds red, 2 diamonds blue and 3 diamonds yellow and the final diamond half green and half white.

9 Student colours 10 beads red, 5 beads blue and 5 beads yellow.

10 a Student colours 5 beads.
b $\frac{3}{4}$ ($\frac{15}{20}$) , 0.75, 75% are white

Extended practice

1

	Item	% offered	Fraction	Number
a	Box of 20 donuts	50%	$\frac{1}{2}$	10
b	Pack of 50 pencils	10%	$\frac{1}{10}$	5
c	Tin of 80 cookies	25%	$\frac{1}{4}$	20
d	Bag of 1000 marbles	1%	$\frac{1}{100}$	10

2 a 50 cm
b 1 metre (100 cm)
c 2 metres (200 cm)

3 Teacher to check. Students could discuss beforehand what they predict will happen. They could also experiment to see what happens if they scale a shape vertically by a different percentage to the horizontal scaling. The reporting could be done orally to a group or on a separate piece of paper.

UNIT 3: Topic 1

Guided practice

1 $150

2 a $50   **b** $75
c $100   **d** $125

3 a $21.50   **b** $43
c $10.75   **d** $107.50

4 a $215.00   **b** $21.50   **c** $193.50

5 $42.50

Independent practice

1 $7.50

2 a 50%, a half, 0.5   **b** $25

3

Description	Quantity	Price per kilogram	Cost
Apples	5 kg	$4.00	$20.00
Pears	5 kg	$1.50	$7.50
Oranges	5 kg	$3.00	$15.00
Bananas	5 kg	$2.00	$10.00
Grapes	2.5 kg	$10.00	$25.00
Total:			$77.50
10% discount if you pay by tomorrow. Discount:			$7.75
Discounted total:			$69.75

4 $30.25

5 Choice 1: Spoons and bowls. 100 spoons + 100 bowls will cost $5.50 plus $22.00 = $27.50, making a total outlay of $97.25. This would generate a profit of $52.75.
Choice 2: Spoons and cups. 100 spoons + 100 cups will cost $5.50 plus $16.50 = $22.00, making a total outlay of $91.75. The profit would therefore be greater ($58.25).

6 The GST is $2 and the total is $22.00

7 $5.00 + $20.00 = $25.00 before GST. GST amount is $2.50 making a total of $27.50

8

Furniture World			
Item	Quantity	Unit price	Cost
Table	1	$120.00	$120.00
Chairs	4	$20.00	$80.00
Price of goods			$200.00
GST (10%)			$20.00
Total:			$220

Furniture For You			
Item	Quantity	Unit price	Cost
Table	1	$130.00	$130.00
Chairs	4	$21.50	$86.00
Total price of goods (including GST)			$216.00

9 a Furniture World: $220 less $22 = $198.
b Furniture For You: $216 less $21.60 = $194.40

Extended practice

1 $82 ($82 plus 10% or $8.20) = $90.20

2 $20

3 Practical activity. Discussion could be held about rounding and on what to do if an amount such as $34 is entered giving a pre-GST total of ($30.9090909).
Students could also be shown that, by clicking and dragging downwards on the + sign at the bottom right corner of cell B2, amounts can be entered in cells A3, A4, and so on.

4 $9.09

UNIT 4: Topic 1

Guided practice

1

Position	1	2	3	4	5	6	7	8	9
Number	1	3	5	7	9	11	13	15	17

2 a

Position	1	2	3	4	5	6	7	8	9
Number	100	98	96	94	92	90	88	86	84

Rule: The numbers decrease by two each time.

b

Position	1	2	3	4	5	6	7	8	9
Number	$\frac{1}{2}$	1	$1\frac{1}{2}$	2	$2\frac{1}{2}$	3	$3\frac{1}{2}$	4	$4\frac{1}{2}$

Rule: The numbers increase by a half each time.

3

Number	12	15
Is it even?	YES ÷ 2	NO, −1, ÷ 2
Answer:	6	7
Is it even?	YES ÷ 2	NO, −1, ÷ 2
Answer:	3	3
Is it even?	NO, −1, ÷ 2	NO, −1, ÷ 2
Answer:	1	1
Is it even?	NO, −1, ÷ 2	NO, −1, ÷ 2
Answer:	0	0

4 a 4

b 6

Independent practice

1 a

Term	1	2	3	4	5	6	7	8	9	10
Number	5	9	13	17	21	25	29	33	37	41

b

Term	1	2	3	4	5	6	7	8	9	10
Number	10	9.5	9	8.5	8	7.5	7	6.5	6	5.5

2 a 0.8, 1, 1.2, 1.4, 1.6, 1.8. Increase by 0.2 each time.

b $3\frac{3}{4}$, $4\frac{1}{2}$, $5\frac{1}{4}$, 6, $6\frac{3}{4}$, $7\frac{1}{2}$. The numbers increase by $\frac{3}{4}$ each time.

3

Number	22 (5 steps)
Is it even?	YES ÷ 2
Answer:	11
Is it even?	NO, −1, ÷ 2
Answer:	5
Is it even?	NO, −1, ÷ 2
Answer:	2
Is it even?	YES ÷ 2
Answer:	0

4 a Step 1: 50 ÷ 2 = 25
Step 2: (25 − 5) ÷ 2 = 10
Step 3: 10 ÷ 2 = 5
Step 4: (5 − 5) ÷ 5 = 0

b Step 1: (125 − 5) ÷ 2 = 60
Step 2: 60 ÷ 2 = 30
Step 3: 30 ÷ 2 = 15
Step 4: (15 − 5) ÷ 2 = 5
Step 5: (5 − 5) ÷ 2 = 0

5 a Increase the number of sticks by 4 for each new diamond.

Number of diamonds	1	2	3	4
Number of sticks	4	8	12	16

b Start with 6 sticks. Increase the number of sticks by 6 for each new hexagon.

Number of hexagons	1	2	3	4
Number of sticks	6	12	18	24

c Start with 5 sticks. Increase the number of sticks by 5 for each new pentagon.

Number of pentagons	1	2	3	4
Number of sticks	5	10	15	20

6 a Start with 4 sticks. Increase the number of sticks by 3 for each new square.

Number of squares	1	2	3	4
Number of sticks	4	7	10	13

b Start with 6 sticks. Increase the number of sticks by 5 for each new hexagon.

Number of hexagons	1	2	3	4
Number of sticks	6	11	16	21

7 Squares: 31, hexagons: 51. (Teachers could ask student to share the strategies they used.)

Extended practice

1 a 1 out of 5   **b** 2
c 20   **d** 200
e 800 (1000 − 200)

2

Number of cars	1	2	3	4	5	6	7	8	9	10
Number of wheels	4	8	12	16	20	24	28	32	36	40

3 a 100   **b** 400
c 1000   **d** 4000

4 a 200 + 2 spares for 50 cars = 202
b 400 + 4 spares for 100 cars = 404
c 1400 + 14 spares for 350 cars = 1414
d 5000 + 50 spares for 1250 cars = 5050

UNIT 4: Topic 2

Guided practice

1 All additions: yes   All subtractions: no
All multiplications: yes   All divisions: no

2 a The answer is the same if you change the order of the numbers for addition and multiplication.

b The answer is not the same if you change the order of the numbers for subtraction and division.

Independent practice

1 Students could be asked to discuss the effective strategies. Easiest solutions are those where rounded sums are found first, e.g.
a 15 + 5 + 17 = 37   **b** 23 + 7 + 19 = 49
c 5 × 2 × 14 = 140   **d** 4 × 25 × 13 = 1300

2 Students could be asked to discuss the problems.
a 10 & 10   **b** 18 & 18   **c** 5 & 5

3 a 2   **b** 3   **c** 4

4 Note: Accept variations using the same numbers, e.g. 26 − 14 = 12 or 72 ÷ 9 = 8.

	Addition and subtraction		Multiplication and division	
	Addition sentence	Subtraction sentence	Multiplication sentence	Division sentence
a	14 + 12 = 26	26 − 12 = 14	9 × 8 = 72	72 ÷ 8 = 9
b	35 + 15 = 50	50 − 15 = 35	25 × 4 = 100	100 ÷ 4 = 25
c	22 + 18 = 40	40 − 18 = 22	15 × 10 = 150	150 ÷ 10 = 15
d	19 + 11 = 30	30 − 11 = 19	20 × 6 = 120	120 ÷ 6 = 20

5 a 4 × 2 = 2 + 6   **b** 18 ÷ 2 = 3 + 6
c 16 ÷ 2 = 2 × 4   **d** 24 − 14 = 3 + 7
e 40 ÷ 2 = 4 × 5   **f** 9 × 2 = 36 ÷ 2
g 2 × 7 = 8 + 6   **h** 50 − 20 = 5 × 6
i 30 ÷ 3 = 100 ÷ 10

6 60 ÷ 5 ÷ 2

7 12 + 2 + 12

8 15 ÷ 4 = 4 ÷ 15

9 Multiple answers possible. Students could be asked to use calculators to check answers. Look for students who use a variety of the four operations to balance the equations.

Extended practice

1

	Problem 1	Problem 2
a	14 − 13 + 7 = 8	14 + 7 − 13 = 8
b	49 − 24 + 25 = 50	25 − 24 + 49 = 50
c	35 − 10 + 25 = 50	35 + 25 − 10 = 50
d	175 − 50 + 25 = 150	175 + 25 − 50 = 150

2 Teachers may wish to ensure that the students understand the order of operations before they complete the activities. ["O" from BODMAS is deliberately missing for Year 5.]

	Problem 1	Problem 2
a	7 + 2 × 3 = 13	(7 + 2) × 3 = 27
b	10 − 8 ÷ 2 = 6	(10 − 8) ÷ 2 = 1
c	15 ÷ 3 + 2 = 7	15 ÷ (3 + 2) = 3
d	10 × 5 + 15 = 65	10 × (5 + 15) = 200

3 Teacher to check, e.g. Because the order of operations means that, in the first problem, 3 is multiplied by 5 first and then the answer is added to 4. In the second problem, 4 is added to 3 first and the sum is multiplied by 5.

4 a Teacher to check, e.g. Because doing 4 × 2 first would mean that Tran lost $4 twice and this did not happen.
b (10 − 4) × 2

5 Teacher to check scenario, but it must suit the sum of 12 and 6 divided by 3.

OXFORD UNIVERSITY PRESS

UNIT 5: Topic 1

Guided practice

1 9 cm

2 a 8 cm   **b** 4 cm   **c** 7 cm

3 a 7 cm 1 mm or 7.1 cm
   b 4 cm 5 mm or 4.5 cm
   c 6 cm 7 mm or 6.7 cm

4 Discuss reasons for tolerance in measuring length with students. Allow +/– 0.1 cm for each line.
   a 3 cm 7 mm or 3.7 cm
   b 6 cm 3 mm or 6.3 cm
   c 9 cm 4 mm or 9.4 cm

Independent practice

1 Teacher to check, e.g. Because it is a rectangle and the opposite sides are the same length.

2 a 18 cm   **b** 12 cm   **c** 16 cm

3 a one   **b** 14 cm (4 × 3.5 cm)

4 Teachers will probably wish to have further discussions about tolerance when measuring perimeter with students. For example, should we allow 4 times +/– 1 mm for each side?
   a 2.5 cm × 2 cm. P = 9 cm
     Number of lines: 2
   b 3.5 cm × 1.5 cm. P = 10 cm
     Number of lines: 2
   c 2.5 cm square. P = 10 cm
     Number of lines: 1
   d 2.5 cm all sides. P = 7.5 cm
     Number of lines: 1

Students will hopefully see that the most time-effective way was to measure two sides of Shapes A and B and one side of C and D.

5

	Centimetres	Millimetres
a	2 cm	20 mm
b	7 cm	70 mm
c	9 cm	90 mm
d	3.5 cm	35 mm
e	7.5 cm	75 mm

6

	Metres	Centimetres
a	2 m	200 cm
b	3 m	300 cm
c	7 m	700 cm
d	$\frac{1}{2}$ m or 0.5 m	500 cm
e	$9\frac{1}{2}$ m	950 cm

7

	Kilometres	Metres
a	2 km	2000 m
b	4 km	4000 m
c	5.5 km	5500 m
d	9.5 km	9500 m
e	8.5 km	8500 m

8 Teacher to check appropriateness of answers. Students could be asked to justify their responses to their peers. Varied answers are possible but likely responses are:
   a centimetres & millimetres
   b centimetres & metres
   c centimetres & millimetres
   d metres & kilometres

9 Allow +/– 4 mm for each shape (at teacher's discretion).
   a 2.2 cm × 1.6 cm. P = 76 mm or 7.6 cm
   b 2.7 cm × 2.3 cm. P = 100 mm or 10 cm
   c 2.9 cm × 1.6 cm. P = 90 mm or 9 cm
   d 1.5 cm × 2 cm × 2.5 cm.
     P = 60 mm or 6 cm

Extended practice

1 & 2 Practical activities. The main aim is for students to practise drawing lines with a reasonable level of accuracy. It is doubtful that 100% accuracy will be obtained and teachers may wish to discuss the reasons for this with students. Set squares could be made available for these tasks.

3 A variety of answers are possible, including 14 cm 5 mm, 14.5 cm, 145 mm and 0.145 m

4 The total length of the line should be 15.5 cm. This could also be written as 155 mm or 15 cm 5 mm.

5 Students should see that, since the shapes are all regular, they need to multiply the given length by the number of sides.
   a 63 mm or 6.3 cm
   b 264 mm or 26.4 cm
   c 114 mm or 11.4 cm
   d 168 mm or 16.8 cm
   e 175 mm or 17.5 cm

UNIT 5: Topic 2

Guided practice

1 a 20 cm²   **b** 25 cm²   **c** 16 cm²
   d 16 cm²   **e** 18 cm²

2 a 8 cm²   **b** 12 cm²   **c** 9 cm²
   d 18 cm²   **e** 12 cm²

Independent practice

1 a 2 rows of 5 cm² = 10 cm²
   b 3 rows of 5 cm² = 15 cm²
   c 3 rows of 7 cm² = 21 cm²
   d 14 cm²   **e** 15 cm²
   f 30 cm²   **g** 25 cm²

2 Students could use centimetre grid overlays.
   a 10 cm²   **b** 6 cm²   **c** 15 cm²
   d 20 cm²   **e** 28 cm²   **f** 16 cm²
   g 36 cm²

3 Students could be asked to share strategies for finding the areas with their peers before, during or after this activity.
   a 12 cm²   **b** 32 cm²

Extended practice

Teachers may wish to discuss the formula for finding the area of a rectangle with students who have demonstrated a complete understanding of the activities on the previous pages.

1 a 5 cm × 3 cm = 15 cm²
   b 4 cm × 2 cm = 8 cm²
   c 3 cm × 3 cm = 9 cm²

2 a A = 4 cm², B = 12 cm², Total = 16 cm²
   b A = 10 cm², B = 12 cm², Total = 22 cm²
   c A = 6 cm², B = 8 cm², C = 6 cm², Total = 20 cm²
   d A = 6 cm², B = 4 cm², C = 6 cm², Total = 16 cm²
   e Student to use own strategy for finding the area. This is likely to be 12 cm² + 2 cm² + 2 cm² = 16 cm² or 8 cm² + 8 cm² = 16 cm².
   f Student to use own strategy for finding the area. This could be 16 cm² + 4 cm² = 20 cm² or 12 cm² + 8 cm² = 20 cm².

UNIT 5: Topic 3

Guided practice

1 a 4 cm³   **b** 2 cm³   **c** 8 cm³
   d 8 cm³   **e** 8 cm³

2 a 600 mL   **b** 2 L
   c 300 mL   **d** 8 L

3 Answers will vary, e.g. a milk carton

Independent practice

1 a 10 cm³   **b** 12 cm³   **c** 20 cm³
   d 16 cm³   **e** 28 cm³

2 Students should by this stage be aware that the volume of a rectangular prism can be found by discovering how many cubes will fit on one layer, and finding multiples of that number (the volume of a single layer multiplied by the total number of layers). In other words, because the number of cubes is the same on the every layer, they are leading towards the formula of V = L × W × H.
   a 16   **b** 16 cm³   **c** 24 cm³

3 a 9   **b** 4   **c** 36 cm³

4 (See note for question 2, above.) Teacher to check, e.g. because the box will hold 2 rows of 4 cubes.

5 a 16 cm³   **b** 24 cm³
   c 32 cm³   **d** 40 cm³

6

	litres	millilitres
a	2 L	2000 mL
b	3 L	3000 mL
c	9 L	9000 mL
d	5.5 L	5500 mL
e	2.5 mL	2500 mL
f	1.25 L	1250 mL
g	3.75 (0) L	3750 mL

7 a 2350 mL, 2 L 400 mL, 2.5 L
   b 0.35 L, 450 mL, $\frac{1}{2}$ L
   c $1\frac{3}{4}$ L, 1.8 L, 1850 mL
   d 20 mL, 200 mL, $\frac{1}{4}$ L

8 D (600 mL)

9 a 1 fruit juice and 1 apple drink
     Amount: **800** mL
   b 2 orange drinks
     Amount: **1500** mL
   c 1 water and 1 apple drink
     Amount: **975** mL. Teacher to decide on an acceptable level of accuracy. Shading should come close to, but below, the 1-litre mark.

Extended practice

1 a 30 cm³

b (See answers to Independent practice, question 2.) Answers will vary, e.g. Because 10 cubes will fit on the bottom layer and there are three layers like that. So, the volume is 10 cm³ × 3 = 30 cm³.

2 a 8 cm³ **b** 36 cm³ **c** 160 cm³
d 100 cm³ **e** 72 cm³ **f** 27 cm³

3 Practical activity. Teachers may wish to use this task for a small or large group activity. It is likely, with normal classroom equipment, that 20 cubes will not displace exactly 20 mL of water. The reasons for this (e.g. inaccuracy of measuring jugs) could be used to promote useful discussion. As an extension activity, if available, a 1000 cm³ cube could be used with a displacement container. This is more likely to displace approximately 1 litre (1000 mL) of water.

UNIT 5: Topic 4

Guided practice

1 a kilograms **b** grams
c tonnes **d** milligrams

2 a

Tonnes	Kilograms
2 t	2000 kg
4 t	4000 kg
1.5 t	1500 kg
3.5 t	3500 kg
1.25 t	1250 kg

b

Kilograms	Grams
2 kg	2000 g
5 kg	5000 g
3.5 kg	3500 g
1.25(0) kg	1250 g
0.5 kg	500 g

c

Grams	Milligrams
5 g	500 mg
3 g	3000 mg
1.5 mg	1500 mg
2.5 g	2500 mg
0.5 g	500 mg

3 a 200 g **b** 600 g
c 1200 g **d** 1900 g

Independent practice

1

	Kilograms and fraction	Kilograms and decimal	Kilograms and grams
a	$1\frac{1}{2}$ kg	1.5 kg	1 kg 500 g
b	$2\frac{1}{4}$ kg	2.25(0) kg	2 kg 250 g
c	$4\frac{3}{4}$ kg	4.75 kg	4 kg 750 g
d	$1\frac{3}{10}$ kg	1.3 kg	1 kg 300 g

2 a 3 kg 500 g, 3.5 kg
b 2 kg 400 g, 2.4 kg
c 4 kg 750 g, 4.75(0) kg
d 1 kg 200 g, 1.2(00) kg

---

3 Answers may vary. Teachers could ask students to justify their responses. Likely answers:
a D or B **b** D
c C **d** A, B or C

4 a **b**

c **d**

5 a Truck B, Truck D, Truck C, Truck A
b Trucks D, C & B
c Trucks B & C
d True (9.05 t)

6 Answers will vary. Look for students who write appropriate responses. For example, apple A appears to be the lightest and the others to be of a similar weight to each other. A simple solution would be to subtract 100 g from 500g and choose masses such as 132 g, 133 g and 135 g for the other three apples.

7 a 4500 kg **b** 350 g
c 15 g **d** 35 kg

8 Yes (total = 1983 kg or 1.983 t)

Extended practice

1 a Blueberry, strawberry, peach, apple, pear, lemon, cabbage, pumpkin
b 724.84 kg **c** 3.165 kg
d Apple **e** 4 (231 g × 4 = 924 g)
f 219.72 g

2 a 39.122 kg **b** 20 (800 ÷ 40 = 20)

3 125 g

UNIT 5: Topic 5

Guided practice

1 a 9:10 am **b** 4:50 pm **c** 11:25 pm
d 1:12 pm **e** 7:19 am **f** 3:47 pm
g 2:22 am

2 Teacher to check clocks and to decide on degree of accuracy for placement of hour hand.
a 8:35 am **b** 6:20 pm
c 11:26 pm **d** 2:47 am

Independent practice

1

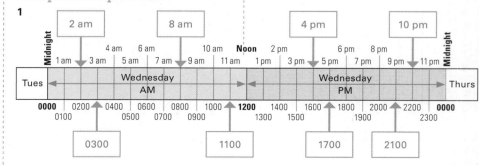

---

2 a 1000 **b** 1530 **c** 1420
d 0711 **e** 2148 **f** 1911
g 0948 **h** 0029

3 Teacher to check. The important thing here is for the student to convert accurately between am/pm and 24-hour times. Teachers may wish to encourage students not to use "o'clock" times.

4 Starting time
2:20

Finishing time
3:05

5 a 3:37 / am/pm / 3:37 pm / 24-hour / 1537
b 10:43 / am/pm / 10:43 pm / 24-hour / 2243
c 7:28 / am/pm / 7:28 am / 24-hour / 0728
d 8:37 / am/pm / 8:37 am / 24-hour / 0837

6 a 10 am **b** 1 pm
c 18 minutes **d** 50 minutes
e 1 hour and 42 minutes (102 minutes)
f Answers may vary, e.g. 1440

7 a 0315 **b** 1515
c 2127 **d** 0927

Extended practice

1 a 23 minutes **b** 2 minutes
c 2 minutes **d** 12 minutes
e 1 hour 50 minutes
f 1755 or 5:55 pm
g 1550

OXFORD UNIVERSITY PRESS